Multispecies Households in the Saian Mountains

Multispecies Households in the Saian Mountains

Ecology at the Russia-Mongolia Border

Edited by
Alex Oehler and Anna Varfolomeeva

LEXINGTON BOOKS
Lanham • Boulder • New York • London

Published by Lexington Books
An imprint of The Rowman & Littlefield Publishing Group, Inc.
4501 Forbes Boulevard, Suite 200, Lanham, Maryland 20706
www.rowman.com

6 Tinworth Street, London SE11 5AL, United Kingdom

British Library Cataloguing in Publication Information Available

Library of Congress Cataloging-in-Publication Data
Names: Oehler, Alex, editor.
Title: Multispecies households in the Saian Mountains : ecology at the Russia-Mongolia border / edited by Alex Oehler and Anna Varfolomeeva.
Description: Lanham : Lexington Books, [2019] | Includes bibliographical references and index.
Identifiers: LCCN 2019043124 (print) | LCCN 2019043125 (ebook) | ISBN 9781793602534 (cloth) | ISBN 9781793602541 (epub)
Subjects: LCSH: Human ecology—Russia Federation—Sayan Mountains. | Animals—Effect of human beings on—Russia (Federation)—Sayan Mountains. | Human beings—Effect of environment on—Russia (Federation)—Sayan Mountains. | Human-animal relationships—Russia (Federation)—Sayan Mountains. | Ecology.
Classification: LCC GF676 .M85 2019 (print) | LCC GF676 (ebook) | DDC 304.20957/5—dc23
LC record available at https://lccn.loc.gov/2019043124
LC ebook record available at https://lccn.loc.gov/2019043125

Contents

List of Figures

Acknowledgments

The idea for this joint intellectual project was first born in the corridors of the Department of Anthropology at the University of Aberdeen in 2016. A number of young scholars focusing on Siberia and the North were then congregating under the direction of David G. Anderson and his European Research Council (ERC) funded Arctic Domus project (No. 295458). This was where the editors of this volume first met in person. The other contributors did not work directly with Arctic Domus, but came together through international research networks, conferences, and social media connections.

The work of Alex Oehler was supported by University of Aberdeen (The North Program under Tim Ingold) from 2012 to 2015, Arctic Domus (ERC) from 2013 to 2017, and a Seed Grant (No. 6005720) from the University of Northern British Columbia in 2018. Much inspiration also came from Tatiana Saburova, lecturer in the Department of History at Indiana University, and the Sixth International Conference of Young Scholars of Siberia in Listvianka, Lake Baikal, in May 2018. Support for workshop attendance came from Indiana University.

Support for the research of Anna Varfolomeeva came from Central European University (CEU) Research Support Grant (2016) and the Fieldwork Grant from CEU Department of Environmental Sciences and Policy (2017). Special thanks to David G. Anderson (University of Aberdeen), to Frank Sejersen and Lars Højer (University of Copenhagen) and to the organizers and participants of the *Siberia: Infrastructure and Environment* workshop in October 2018 in Berlin for their helpful comments and suggestions.

Individual authors acknowledge their funding sources and support in the first endnote of their respective chapters. Collectively, the authors of this volume would like to thank the communities into which they were so warmly received. Special thanks go to the local administrations and community members of Alygdzher, Nizhneudinskii District, Irkutskaia Oblast', Tora-Khem and Ader-Kezhig, Todzhinskii District, Republic of Tyva, Ak-Erik, Tes-Khemskii District, Republic of Tyva, Orlik and Sorok, Okinskii District, Republic of Buriatia, and Gurvansaikhan, Khövsgöl Province, Mongolia.

Introduction

On Making Home Together

Alex Oehler

This volume brings together new ethnographic perspectives from mountainous South Siberia and northern Mongolia. As an international group of emerging and established anthropologists of Inner Asia, we seek to combine in this book fresh insight on ethnohistorical and contemporary Indigenous human-animal relations. We are interested in how the local residents of alpine and subalpine Inner Asia relate to nonhuman sentient others under changing environmental and political circumstances. Our studies are predominantly set in the eastern Saian Mountain Range of south central Siberia and in northernmost Mongolia, a region that is widely known in ethnographic and anthropological literatures as the birthplace of Eurasian reindeer domestication. While reindeer play an important role in this book, it is our aim to transcend this classical focus by introducing a variety of other species that co-constitute South Siberian Mountain households.

Our focus areas stretch along both sides of the border between Mongolia and the Russian Federation, presenting a unique transition zone between the Mongolian plateau, known for its high altitude steppe and widespread pastoralism, and the Siberian taiga, a mountainous forested landscape inhabited by various Indigenous hunting societies. This cultural, linguistic, and ecological nexus forms a rich setting for a qualitative study of encounters between species, cultural practices, and unique ways of knowing. In each of our contributions, we investigate how multispecies encounters constitute nuanced articulations of the household as a concept of social belonging.

FIVE FOCUS AREAS

Each of the chapters is based on fieldwork conducted in one of five areas, located along the border between Russia and Mongolia (see figure 0.1). Starting in the west, Aivaras Jefanovas and Tayana Mongush-Arakchaa discuss their independent work with Tozhu reindeer herders in Todzhinskii District, located in the eastern part of the Republic of Tyva. The Russian census of 2010 registered 1,850 Tozhus (Federal Service of Government Statistics [FSGS]). The majority of Tozhus, who are often referred to as Eastern Tyvinians or Tyvinian-Todzhins [Rus. Tuvintsy-Todzhintsy], speak an eastern dialect of the Tyvan language, which belongs to the Turkic language family. Both authors follow in the footsteps of recent anthropological work (Stépanoff 2012, 2011; Donahoe 2006, 2004, 2011) that has contributed to a growing ethnographic canon for this region. The herders and hunters with whom the authors in this book worked were at the time following their herds deep into the mountainous taiga. For this reason, the field sites of each researcher cannot be pinpointed to permanent settlements, such as the regional center of Tora-Khem or the settlement of Biy-Khem, although herders would regularly return to these villages.

To contrast the emphasis on reindeer in the households of eastern Tyva, Victoria Peemot contributes a study from southern Tyva, where horse-human bonds dominate. The district of Tes-Khem serves well as an example of pastoral transhumance between forested verticality and the grassy horizontal, given its position at the intersection of the Tangdy-Uula [Eng. Tannu-Ola] range and the steppe lands of the Uvs Lake Basin. As is true of the other

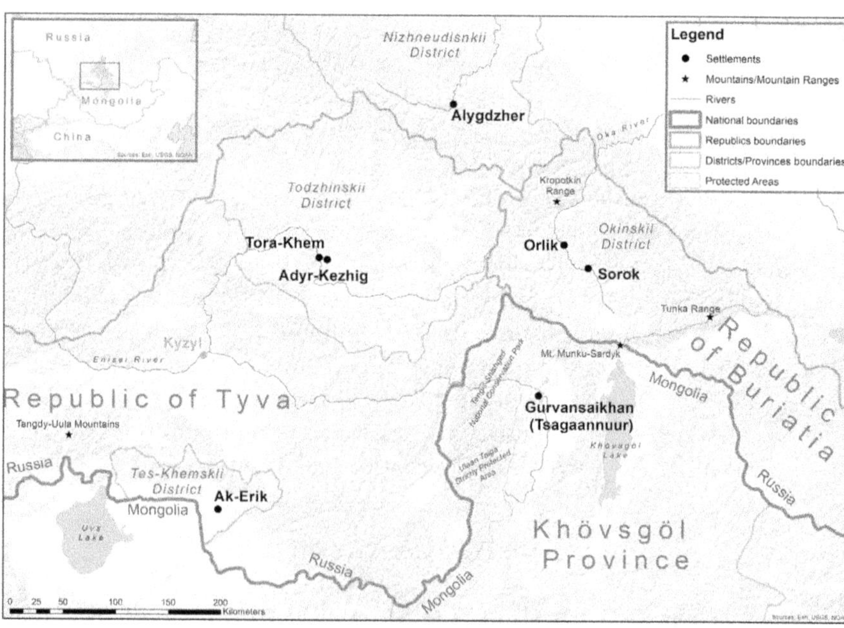

Figure 0.1 The Saian Cross with the Authors' Fieldsites. *Source*: 2019 A. Kvasha.

regions discussed in this book, villages and regional centers often serve as starting points for seasonal movements. In many cases, permanent settlements are home to the families of individual herders and hunters. In other cases, the population has split into families who live in the settlements full time, while others maintain herds in a pattern of transhumance, operating from mobile encampments in the taiga.[1]

Western interest in the people of eastern Tyva was first sparked in connection with questions about the origins of reindeer domestication. The whole Altai-Saian region was then the focus of debates that have since been summarized more than once for an English readership (e.g., Whitaker 1981; Mirov 1945; Zolotarev and Levin 1940). Diffusionist scholars argued, reindeer domestication had spread from the Altai-Saian to the rest of Eurasia, including Scandinavia (Aronsson 1991; Laufer 1974; Hatt 1918; Sirelius 1916). Evolutionist scholars, in contrast, advocated multiple independent origin points (Storli 1996; Mulk 1994; Wiklund 1918). Few of these scholars had conducted firsthand research in Siberia. Among the first Western visitors to eastern Tyva were British explorer Douglas Caruthers (1914) and Norwegian zoologist Ørjan Olsen (1915). Both men visited with Tozhus of eastern Tyva over the course of three summer months in 1910 and 1914, respectively. The father of this region's ethnography and archaeology, however, remains ethnohistorian and archaeologist Sevian Israilovich Vainshtein (1926–2008), who also offered a comprehensive theory for reindeer domestication (1970), but whose extensive work has been translated into English only partially (1980).

Further to the northeast of the Tozhu region—and on the other side of the Bolshoi Saian Range in Irkutskaia Oblast'—authors Konstantin B. Klokov, Alex Oehler, and Igor V. Rassadin describe their encounters with Tofa hunter-herders. Numbering 762, according to the 2010 census (FSGS), Tofas, formerly also known as Tofalars or Karagas, make their ancestral home in Tofalaria, a region located in Nizhneudinsk District of Irkutskaia Oblast'. Contemporary Tofas, who are related to their southwestern Tozhu neighbors, no longer speak Tofa, although efforts are being made to revive the ancestral language, which—like that of Tozhus—belongs to the sub-branch of Siberian Turkic languages. The majority of Tofas reside in three settlements: Alygdzher, Nerkha, and Verkhniaia Gutara. Few Western anthropologists have conducted ethnographic work in this area, and those who have (e.g., Donahoe 2004) follow in the footsteps of their Russian colleagues (Krivonogov 2008; Rassadin 2005; Mel'nikova 1994). Unlike Tora-Khem, which can be accessed via gravel road and by river year-round, Tofalaria's administrative center of Alygdzher is accessible only by helicopter in summer or by ice road in winter. The majority of Tofa households continue to rely on annual hunting for their subsistence, and many still depend on reindeer for transportation in this pursuit.

Tofalaria has been the focus of several pre-revolutionary expeditions, including one by renowned Finnish scholar Matthias Alexander Castrén (1856), who visited the Tofas on his travels from Kansk to Nizhneudinsk in 1848. Castrén was followed by Russian statesman Iulii I. Shtubendorf (1858), and then by well-known Irkutsk publisher Ivan I. Popov (1879). Linguistic interest in Tofas and Tofa folklore did not, however, end with Castrén. In 1890, German-Russian turkologist Vasilii V. Radlov (1907), along with his student and colleague, Khakas-Russian turkologist Nikolai F. Katanov (1891), visited Tofa encampments in the winter of 1890. A more anthropological approach was later taken by Russian ethnographer and folklorist Viktor N. Vasilev (1910) from the Museum of Anthropology and Ethnography at Saint Petersburg, who spent two months with the Tofas in 1908. In spite of these numerous visits, the bulk of our knowledge about pre-Soviet era Tofa life ways comes from the detailed work of Swiss-Russian anthropologist and archaeologist Bernhard Eduardovich Petri (1927a, 1928) who was interested in many of the details of Tofa human-animal relations.

Southeast of Tofalaria, on the southern side of the Kropotkin Range, we enter Okinskii District, the westernmost District of the Republic of Buriatia. This region is locally known as Akha or Oka, named so after the Oka River. Protected from the south by the steep Tunka Peaks, and surrounded by mountains on all other sides, Oka has remained one of the least studied regions of Buriatia. It is populated by Oka-Soiot and Oka-Buriat clans who share a history of extensive intermarriage. Authors Anna Varfolomeeva and Alex Oehler focus primarily on Indigenous herders and hunters who stress a sense of Soiot identity, and who reside either in the District's center of Orlik, or within the traditional use area of Soiotiia, a region that falls under the administration of the village of Sorok. Sorok is locally also known as "the center of Soiot culture and heritage." Having lost their ancestral Turkic language, Soiots now speak Buriat in the home, yet 3,608 residents declared themselves Soiot in the 2010 census (FSGS). Although reindeer were essential to Soiot hunting in the past, their focus has shifted to yak breeding and other pursuits.

In spite of its close proximity to Mongolia and Tyva, Oka received very few visits from scholars and explorers prior to 1920. In fact, famous Russian anarchist Pëtr Alekseevich Kropotkin (1867)—in the days of his earlier career as a geologist—was among the very few outsiders who rode through this region, leaving behind a record of his encounters. What little ethnographic record we have for Oka, and specifically for Oka-Soiots, comes primarily from B. E. Petri (1927b) who, following his work in Tofalaria, conducted a single expedition to Oka in 1926. Commissioned by the Northern Committee [Rus. Komitet Severa], he managed to visit every major Soiot community at the time. After Petri, the only concerted ethnographic and ethnohistorical research on Soiots was conducted by anthropologist Larisa Pavlinskaia (2002), who played an instrumental role in reinstating the rights of Soiots as an Indigenous minority in the Russian Federation. Her work later inspired some Western forays (e.g., Donahoe 2011; Oehler 2018).

Following the Irkut River to the base of Mt. Munku Sardyk (3491m), we find ourselves on the border to northernmost Mongolia. On the south side of the sacred mountain, we enter the Tengis-Shishged National Conservation Park, as well as Ulaan Taiga—the ancestral migratory grounds of Dukhas, known in Mongolia as Tsaatans of Khövsgöl Aimag. Approximately 200 reindeer herding Dukhas are estimated to live in the taiga, while about 300 reside in the village of Gurvansaikhan (Kristensen 2015, 13). Like their related Tozhu, Tofa, and Soiot neighbors, the Dukhas described by authors Benedikte Møller Kristensen, Selcen Küçüküstel, and Nicolas Rasiulis share a dialect of Tyvan, which Dukha elders are still fluent in. Especially in recent years, the Dukhas have received by far the greatest international media attention of all the Indigenous populations discussed in this book. While some of this publicity may have to do with the recent upheaval surrounding the loss of Dukha subsistence rights, which has resulted from the declaration of their ancestral lands as a protected national park, it is more likely that public interest has been sparked by the broad visual depiction of their lifestyle on the Internet and in documentary films. The Dukhas are one of the few nomadic peoples who still reside in conical lodges year-round, rendering their encampments a prime destination for eco and ethnotourism.

All this attention, however, has not always resulted in an improved understanding of the nuanced circumstances that define contemporary and historical Dukha lives. Neither has it necessarily enriched our anthropological understanding of more-than-human Dukha households. Kristensen (2016, 2004), Küçüküstel (2018, 2013), and Rasiulis' (2016) previous work with Dukha households has already strengthened an otherwise slender body of Dukha ethnography. This is not to say that anthropologists (O'Brian & Surovell 2017; Inamura 2005), archaeologists (e.g., Haas, et al. 2018; Haas 2019), ethnohistorians (Wheeler 2000), and social geographers (e.g., Endres 2015, 2012; Badamkhatan 1996), and others have not conducted short or long-term research with nomadic Dukha households in recent years.

THEORETICAL STARTING POINTS

As a writing collective, and following previous work (e.g., Fijn 2011), we believe there is a sustained need for multispecies ethnography that transcends wild-tame dichotomies in southern Siberia and Mongolia. Transcending the single species focus on reindeer in much of the existing literature on the region, this book argues for the importance of a multispecies approach to the study of Indigenous households in the eastern Saians. While reindeer herding and hunting undoubtedly form a collective heritage for the peoples of this area, the reassessment of existing and nascent ethnographic materials calls for a renewed focus on the role of other species in these mountainous households. In this book, we argue for a varied and nuanced interdependence of species as we present carefully adapted seasonal fisheries, regionally specific horse and dog breeding techniques, and in some places more recently introduced species, such as yak. As significant points of connection between humans and animals, these multispecies relations are embedded in a cosmology connecting domestic and semi-domestic animals with species outside the human-managed household.

We find that the relationship between human and other households (often spirit-governed households, embedded in the landscape) is best articulated beyond idioms of dominance or mutualism. At the heart of the book lies the idea of the human household as something that extends beyond the confines of pasture, or of what can be "owned" by people. This is made evident in descriptions of human interactions with animals, plants, and rocks known to belong to the households of spirits. While the adjacency of human and spirit households allows for social and economic transactions, it also enables creatures of various origins to permeate the territorial boundaries of their owners' households. The porous nature of these zones then calls for a re-conceptualization of the domus, or home. Drawing on recent thinking about architectures of domestication, we follow an "architecture of relationships" (Anderson et al. 2017), rooted in the idiom of transspecies familiarity (Takakura 2010), rather than in domination of one entity over another, or in the necessity for mutualism in human-animal relations.

We can see a promising future in researching home places as nexuses of multispecies intentions. For this reason, in the chapters that follow, we interrogate ethnographic situations in which multiple species co-create living spaces, approaching home and household as a nexus of intentions (Gell 1996). Shared living and dying spaces here become a sort of contact zone (Clifford 1997) that is found useful or habitable from the perspective of more than one species. By building on architectures of relationships and nexuses of intentionality, we are able to address material resources like meat, milk, jade, animal tack, and so on as semiotic components in landscapes inhabited by diverse sentient beings. While not attempting to offer a panacea for the challenges we find ourselves confronted with in the Anthropocene, we do seek to provide in this book inspiring material for thinking through the turbulences and unpredictabilities of reconfiguring our relations (Haraway 2016) to the earth and its inhabitants.

IDEAS TO THINK WITH

To substantiate the three aforementioned arguments, our chapters address—sometimes more directly, other times more indirectly—five main topics. Firstly, we are concerned with domestication relationships, which we loosely define as dynamic animal-human relations in which the life experiences of all participants are subject to change resulting from an ongoing negotiation of more-than-human intention. There has been increased interest in the topic of animal domestication in the Social Sciences and Humanities, with many conferences and seminars critically reassessing the concept of domestication, and especially the role of humans in this process. The chapters in this book pick up on some of these debates, contributing valuable case studies that challenge conventional polarities. An example is Mongush-Arakchaa's chapter in which she outlines the ambivalent nature of Tozhu reindeer, horses, and dogs, while showing how wolves inform human hunting schemes.

Secondly, we are interested in more nuanced notions of the home. What we have in mind are home places or locations in the landscape where the paths of humans and other beings converge. By focusing on broader criteria of what might constitute a home, we are able to rethink previously imagined boundaries between human and other animal spaces. Rethinking home places falls in line with the interests of many of our colleagues in human-animal studies, whether they focus on urban or rural (Holmberg 2018; Ziegler 2018), air space (Schroer 2018; Fageraas 2016), or aquatic environments (Lien 2015). For our observations from Inner Asia, we approach home places as locations in which sentient beings invoke artificial and natural features to advance their own or each other's livelihoods through relational architectures. A good example of this is found in Peemot's chapter, which examines the ways in which boundaries in the landscape articulate different notions of home.

A third point of departure lies in the communicative capacity of materials. Much of the literature on animal domestication has framed material implements, such as ropes, lassos, fishing nets, fences, or saddles as implements of human control and domination. We approach materiality from the angle of its capacity for non-vocal communication between sentient beings. While communication may be coercive at times, it does not have to be so, and neither does it have to be unidirectional. This multidirectional capacity for communication through materials is not limited to humans and animals, but also includes spirit entities in all of our research locations. The contributions by Mongush-Arakchaa, Peemot, and Oehler and Rassadin particularly examine such interspecies materialities, whether they are found in association with dogs, horses, or reindeer.

A fourth point of interest to us is cosmology, and more specifically the ways in which people perceive sentience. Scientific scholarship has made use of many definitions of sentience over the years. Yet, we are less interested in nomothetic or objective definitions of sentience. Rather, we seek out accounts of lived Indigenous experiences of sentience in other beings, including landscapes and specific land features. For this reason, our contributions go beyond even some of the more inclusive definitions of sentience currently in use in the sciences. The chapters by Küç̌ükü̆stel, Møller Kristensen, and Rasiulis on Dukhas, and the one by Varfolomeeva on Soiots specifically address these concerns.

Our final aim of inquiry constitutes rhythm and speed in mountainous multispecies households. Although scholars have lent much attention to the intricate demands put upon human householders in the management of diverse species with various feeding, resting, and reproductive cycles, relatively little attention has been paid to how the speed and rhythm of each species is perceived by householders, and how awareness of pace is taken into account when species meet. Various aspects of this concern are addressed across all chapters, while Oehler addresses the question of rhythm more specifically in the final chapter.

FROM DOGS TO YAK

Following this introduction, Tayana Mongush-Arakchaa opens up part I of the book with an examination of Tozhu human-dog relations. In eastern Tyva, the story of early taiga hunters begins with a remarkable achievement—the domestication of canids. With the aid of reindeer and dogs, taiga dwellers developed a successful hunting style by mimicking the hunting tactics of wolves. According to the author, Tozhu human-dog relations are patterned after a wolf-pack approach, in which the Tozhu have learned to use the desirable features of their domesticated animals. The wolf-pack approach is effective in that it helps us better understand herder-hunters' relations with their dogs, particularly as the concept is extended to include other species living alongside humans, such as reindeer and horses.

Aivaras Jefanovas further examines hunter-predator relations in the Tozhu region in chapter 2. On the one hand, he relates how predators are considered dangerous pests; rivals with whom there is a constant struggle for reindeer. On the other hand, he shows how the Tozhus have great respect for predators as intelligent and sensitive beings. Jefanovas goes further, drawing attention to the ways in which tactics known from the human-predator interface come to inform human social relations, breaking down the divide between human and more-than-human sociality. By way of this conflation, the author points to the situational nature of local decision making, which seems to be neither aggressive nor cooperative, but a matter of constant adjustment reflecting a land-based cosmology.

In chapter 3, Victoria Peemot explores the autonomy and hierarchy of horses in multispecies households of South Tyva. The horses she describes share their world with humans and nonhumans in an eco-social landscape shaped by the borderlands between boreal forests and steppe lands. Peemot investigates the different symbolic and physical spaces occupied by horses in homes, encampments, and the open country. She points to the "kodan" territory of the larger encampment and their association with the open country, arguing that the differences between these spaces reflect herders' understandings of horses as simultaneously close to humans and to wilderness. This "not-quite-livestock" placement allows the horse a unique position of autonomy.

Konstantin Klokov opens part II of the book by taking us to Tofalaria, the homeland of Tofas northeast of Tyva. Focusing on their contemporary economy, which remains centered on hunting, he describes contemporary and historical Tofa relations with wild and domestic species, including reindeer, dogs, various furbearers, Manchurian wapiti, bears, and wolves. He examines a series of cultural practices embedded within these relationships, among them methods of training reindeer for riding, dairying, and as hunting companions, as well as selecting and training hunting dogs. Drawing on Tofa terminology, Klokov shows how this multiplicity of species is perceived as a common "family" at home in a shared taiga environment.

Continuing on with Tofalaria in chapter 5, Alex Oehler and Igor Rassadin pick up on Klokov's introduction, as they focus more specifically on the understudied role of horses in Tofa households. Looking back to historical records of regional Russian settler use of horses, as well as to horse trade with Oka-Buriats, the authors reconstruct the seasonal role of equines in this mountainous reindeer economy. In so doing, Rassadin and Oehler draw on material culture and animal-related cosmology that has been shared between reindeer and equines, while analyzing the inseparability of current and historical transhumance patterns for both species.

In chapter 6, Alex Oehler draws on the insights of Tofa reindeer herder Sergey Tulaev, as well as on his own ethnographic observations from Alygdzher's reindeer camp on the Dugol'ma river. The aim is to provide a more detailed overview of how Soviet changes in herding continue to influence current breeding and herding techniques in a place believed by some to have lost its intimate relationship with reindeer. Oehler is particularly interested in the idiom of curiosity as an integral yet problematic constituent of local notions of domestication, and to the nuanced measures taken by herders in preparing and managing reindeer calving grounds as places in the taiga that are conducive to the emergence of vulnerable lives.

Part III sets out with Nicolas Rasiulis, looking at the dynamic composition of Dukha households, which are comprised of humans alongside sympathetic albeit independently minded animals and spirits. Rasiulis pays special attention to the improvisatory and creative qualities of actions and interactions of household dwellers within and beyond the home. He is particularly interested in how such actions help fashion suitable short- and long-term domestic spaces within a sentient, labile, and bountiful, yet often dangerous homeland at a time of rapid socio-economic change.

Rasiulis' reflections are followed by Selcen Küçüküstel, in chapter 8, with an exploration of how Dukha herders and hunters interact with their environment. Küçüküstel is interested in how backcountry navigation relies on delicate relations between people, animals, ancestors, and spirits. Since each household has various spirits that reside with the people, and because the taiga is populated by spirits with different personalities, people rely on detailed spiritual mind maps as they move through their homeland. Furthermore, she discusses how people interact with these spirits, while looking at some of the reasons they decide to remain in the taiga despite contemporary societal changes.

Then, in chapter 9, Benedikte Møller Kristensen explores the multifaceted agency of meat and milk, obtained from wild game and domesticated reindeer of Dukha households. In particular, she focuses on how these products are felt to engage with the life of the household in complex ways, and how the Dukhas attempt to direct and control the powers of these products. Drawing on Derrida's concept of "hauntology" the chapter reveals how the agency of animal products consists in the ability to draw out the accumulated powers of the history of past kin who continue to shape the lives of living members of the household.

Part IV—the final part of this book—begins with Anna Varfolomeeva's study of illegal jade extraction as a part of the local cosmology of Oka in Buriatia. She examines the close connections of Soiots and Oka Buriats with the land, and how these relations impact their perceptions of jade mining. As various parts of local landscape, including subsurface resources, are seen as animated and governed by spirit masters, illegal extraction of jade becomes not only a business activity but also a spiritual journey. Varfolomeeva analyzes complex perceptions of jade mining in Oka, including their parallels with other economic activities such as hunting.

This is followed, in chapter 11, by Alex Oehler's discussion of the challenges of reconciling divergent species' life-paces within Soiot households. He examines the ruptures that occur where households are no longer able to bridge the different requirements for pace and range in certain species, especially in contexts where cultural blending has introduced new species, dictating changes in the speed, frequency, and altitude of household movements. Reindeer, yak, dairy cattle, wolves, fish, and horses are drawn into the discussion to illustrate relations between occupation, feed availability, and migrational distance and speed of movement.

Finally, Varfolomeeva concludes the book with a series of reflections on some of the core themes brought forth by our contributors, while pointing to future directions for qualitative research in this unique region of the world, which for many of the authors has become a new home place.

NOTE

1. Even in the latter case, children who are of school age accompany their parents only during the holidays, staying with relatives in the village while school is in session.

REFERENCES

Anderson, David G., Jan Peter Laurens Loovers, Sara Asu Schroer, and Robert P. Wishart. 2017. "Architectures of Domestication: On Emplacing Human-Animal Relations in the North." *Journal of the Royal Anthropological Institute* 23, no. 2: 398–416. doi:10.1111/1467-9655.12613_1.

Aronsson, Kjell-Åke. 1991. "Forest Reindeer Herding A.D. 1–1800. An Archaeological and Palaeoecological Study in Northern Sweden." PhD dissertation, University of Umeå.

Badamkhatan, S. 1996. "Tuva Tsaatan." In *Mongol Ulsiyn Ugsaatny Züi*, edited by S. Badamkhatan, vol. 3. Ulaanbaatar: Sukhbaatar Press.

Carruthers, Douglas, and Jack Humphrey Miller. 1914. *Unknown Mongolia: A Record of Travel and Exploration in Northwest Mongolia and Dzungaria*, vols. 1 & 2. London: Hutchinson & Company.

Castrén, Matthias, Alexander. 1856. Reiseberichte und Briefe aus den Jahren 1845–1849. Saint Petersburg: Anton Schiefer.

Clifford, James. 1997. *Museums as Contact Zones*. Cambridge: Harvard University Press.

Donahoe, Brian. 2004. "A Line in the Sayans: History and Divergent Perceptions of Property Among the Tozhu and Tofa of South Siberia." PhD diss., Indiana University.

———. 2006. "Who Owns the Taiga? Inclusive vs. Exclusive Senses of Property Among the Tozhu and Tofa of Southern Siberia." *Sibirica* 5, no. 1: 87–116. doi:10.3167/136173606780265306.

———. 2011. "On the Creation of Indigenous Subjects in the Russian Federation." *Citizenship Studies* 15, no. 3–4: 397–417. doi:10.1080/13621025.2011.564803

Endres, Jürg. 2012. "Die Dukha: Rentierhalter und Jäger der Mongolei Zwischen Subsistenzjagd, 'Wilderei' und Gemeindeorientiertem Ressourcemanagement." In *Naturschutz und Freizeitgesellschaft: Minderheiten im Umgang mit der Natur*, edited by K. Hammerlich, M. Klein, and M. Romich, 147–168. Sankt Augustin: Academia Verlag.

———. 2015. *Rentierhalter. Jäger. Wilderer? Praxis, Wandel und Verwundbarkeit bei den Dukha und den Tozhu im mongolisch—russischen Grenzraum*. Stuttgart: Franz Steiner Verlag.

Fageraas, Knut. 2016. "Housing Eiders—Making Heritage: The Changing Context of the Human-Eider Relationship in the Vega Archipelago, Norway." In *Animal*

Housing and Human-Animal Relations: Politics, Practices and Infrastructures, edited by K. Bjørkdahl and T. Druglitrø, 82–99. Oxon: Routledge.

Fijn, Natasha. 2011. *Living with Herds: Human-Animal Coexistence in Mongolia.* Cambridge: Cambridge University Press.

FSGS (Federal Service of Government Statistics). *Natsional'nyi Sostav Naselenia* [National Population Composition], 2010. Accessed July 25, 2019. http://www.gks.ru/free_doc/new_site/perepis2010/croc/documents/vol4/pub-04-01.pdf

Gell, Alfred. 1996. "Vogel's Net : Traps As Artworks And Artworks As Traps." 1 (1): 15–38. doi:10.1177/135918359600100102.

Haas, Randall, Todd Surovell, and Matthew O'Brien. 2018. "Occupancy and the Use of Household Space Among The Dukha." *Ethnoarchaeology* 10, no. 1: 1–15. doi: 10.1080/19442890.2018.1440510.

———. 2019. "Dukha Mobility in a Constructed Environment: Past Camp Use Predicts Future Use in the Mongolian Taiga." *American Antiquity* 84, no. 2: 215–233. doi:10.1017/aaq.2018.88

Hatt, Gudmund. 1918. "Rensdyrnomadismens Elementer." *Geografisk Tidskrift*, no. 24: 241–269.

Holmberg, Tora. 2018. "'Moving Quietly in the Shadows': On Feral Feeding in Kolkata." In *Animal Places: Lively Cartographies of Human-Animal Relations*, edited by Jacob Bull, Tora Holmberg, and Cecilia Åsberg, 33–50. London: Routledge.

Inamura, Tetsuya. 2005. "The Transformation of the Community of Tsaatan Reindeer Herders in Mongolia and Their Relationships with the Outside World." In *Pastoralists and Their Neighbors in Asia and Africa*, edited by K. Ikeya and E. Fratkin, 123–152. Osaka: National Museum of Ethnology.

Katanov, Nikolai F. 1891. "Poezdka k Karagasam v 1890 godu. [A Voyage to the Karagases]" *ZRGO po Otdelu Etnografii* [Records of the Russian Geographical Society, Department of Ethnography] 17(2).

Kristensen, Benedikte Møller. 2004. "The Living Landscape of Knowledge: An Analysis of Shamanism Among the Duha Tuvinians of Northern Mongolia." Institut for Antropologi, Københavns Universitet. http://www.anthrobase.com/Txt/K/Kristensen_B_02.htm.

———. 2015. "Returning to the Forest: Shamanism, Landscape and History Among the Duha of Northern Mongolia." PhD diss., University of Copenhagen.

———. 2016. "The Proper Funeral: Death, Landscape and Power Among the Duha Tuvinians of Northern Mongolia." In *Governing the Dead*, edited by Finn Stepputat, 35–52. Manchester: Hive. doi:10.7765/9781847799111.00012

Krivonogov, Viktor P. 2008. *Tofalary: Tri Shaga v Budushchee* [Tofalars: Three Steps Into the Future]. Krasnoiarsk: Krasnoiarsk State Pedagogical University.

Kropotkin, Petr A. 1867. "Poezdka v Okinskii Karaul" [Voyage into the Oka Fore-Post]. *Zapiski Sibirskogo Otdelenia Russkogo Geograficheskogo Obshchestva*, no. 9: no pages. Irkutsk: IRGO. https://ru.m.wikisource.org/wiki/Поездка_в_Окинский_караул_(Кропоткин)/Предисловие. Accessed: Nov. 13, 2019.

Küçüküstel, Selcen. 2013. "The Dukha and How They Perceive the Environment–Maintaining Good Relations with Land Spirits." MA thesis, Yeditepe University, Istanbul.

———. 2018. "Living with Reindeer and Hunting Among Spirits: The Dukha of South Siberia." PhD diss., Humbold-Universität zu Berlin.

Laufer, Berthold. 1974. "The Reindeer and its Domestication." *Memoirs of the American Anthropological Association* 4, no. 2: 91–147.

Lien, Marianne Elisabeth. 2015. *Becoming Salmon: Aquaculture and the Domestication of a Fish*. Berkeley, CA: University of California Press.

Mirov, N. T. 1945. "Notes on the Domestication of Reindeer." *American Anthropologist* 47, no. 3: 393–408.

Mulk, I. M. 1994. "Sirkas—ett Samiskt Fångstsamhälle i Förändring Kr. f.-1600 e. Kr." PhD dissertation, Umeå University.

O'Brien, Matthew J., and Todd A. Surovell. 2017. "Spatial Expression of Kinship Among the Dukha Reindeer Herders Of Northern Mongolia." *Arctic Anthropology* 54, no. 1: 110–119. doi:10.3368/aa.54.1.110.

Oehler, Alexander. 2018. "Hunters in Their Own Right: Perspectival Sharing in Soiot Hunter's and their Dogs." In *Dogs in the North*, edited by Robert J. Losey, Robert P. Wishart, and Jan Peter Laurens Loovers, 28–44. London: Routledge.

Olsen, Ørjan. 1915. *Et Primitivt Folk: de Mongolske Rennomader*. Oslo: Cappeln.

Pavlinskaia, Larisa. 2002. [Nomads of the Blue Mountains. The Fate of Traditional Culture of the Peoples in the Eastern Saians in the Context of their Interactions with Modernity] St. Petersburg: Evropeiskii Dom.

Petri, Bernhard. 1927a. *Olenevodstvo u Karagas* [Reindeer Herding of the Karagasses]. Irkutsk: Izvestiia Biologo-Geograficheskogo Nauchno-Issledovatel'skogo Instituta pri Gosudarstvennom Irkutskom Universitete [Communications of the Bio-Geographical Scientific Research Institute of the State University of Irkutsk].

———. 1927b. *Etnograficheskie Issledovania sredi Malykh Narodov v Vostochnykh Saianakh: Predvaritel'nye Dannye* [Ethnographic Researches Among the Small Numbered Peoples of the Eastern Saians: Preliminary Findings]. Irkutsk.

———. 1928. "*Promysly Karagas* [Subsistence Strategies of Karagasses]." *Vlast' Truda* [The Power of Labor], 52.

Popov, I. I. 1879. "Zapiska of Karagassakh [A note on the Karagas]." In *O Pervykh Priemakh Fizicheskago Vospitaniia Detei u Razdnykh Narodnostei, Naseliaiushikh Irkutskuiu Guberniu* [Methods of Early Childhood Physical Education among Diverse Nationalities Residing within Irkutsk Province]. Irkutsk: Irkutskii Gubernskii Statisticheskii Komitet [Statistical Committee of the Irkutsk Governorate].

Radlov, Vasilii V. 1907. *Proben der Volksliteratur der Türkischen Stämme: Mundarten der Urianchaier (Sojonen), Abakan-Tataren und Karagassen*. Saint Petersburg.

Rasiulis, Nicolas. 2016. "Freestyle Bearing: Work, Play, and Synergy in the Practice of Everyday Life Among Mongolian Reindeer Pastoralists." MA thesis, University of Ottawa.

Rassadin, Igor, V. 2005. *Khoziaistvo, Byt i Kul'tura Tofalarov*. Ulan-Ude: IMBIT.

Schroer, Sara Asu. 2018. "Breeding with Birds of Prey: Intimate Encounters." In *Domestication Gone Wild: Politics and Practices of Multispecies Relations*, edited by Heather Anne Swanson, Marianne Elisabeth Lien, and Gro B. Ween, 33–49. Durham: Duke University Press.

Sirelius, U. T. 1916. "Uber die Art und Zeit der Zähmung des Rentiers." *Journal de La Societé Finno- Ougrienne*, no. 33: 4–10.
Stépanoff, Charles. 2011. "Saillances et Essences. Le Traitement Cognitif de la Singularité Chez les Éleveurs de Rennes Tožu (Sibérie Méridionale)." *L'Homme*, no. 200: 175–202.
———. 2012. "Human-Animal 'Joint Commitment' in a Reindeer Herding System." *HAU: Journal of Ethnographic Theory* 2, no. 2: 287–312.doi:10.14318/hau2.2.015
Storli, Inger. 1996. "On the Historiography of Sami Reindeer Pastoralism." *Acta Borealia* 13, no. 1: 81–115.
Shtubendorf, Iulii I. 1858. *O Karagasakh*. Pp. 1–18.
Takakura, Hiroki. 2010. "Arctic Pastoralism in a Subsistence Continuum: A Strategy for Differentiating Familiarity with Animals." In *Good to Eat, Good to Live With: Nomads and Animals in Northern Eurasia and Africa*, edited by Florian Stammler and Hiroki Takakura, 21–42. Sendai: Center for Northeast Asian Studies.
Vainshtein, Sevyan I. 1970. "Problema Proizkhozhdenia Olenevodstva v Evrazii [The Problem of the Origin of Reindeer Domestication in Eurasia]." *Sovietskaia Etnografia* [Soviet Ethnography], no. 1: 37–52.
Vasilev, Viktor N. 1910. "Kratkii Otsherk Byta Karagasov [A Brief Overview of Karagas Subsistence]." [Ethnographic Review], 1–2, no. 84–85: 46–76.
———. 1980. *Nomads of South Siberia*. Cambridge: Cambridge University Press.
Wheeler, W. Alan. 2000. "Lords of the Mongolian Taiga: An Ethnohistory of Dukha Reindeer Herders." MA thesis, Indiana University.
Whitaker, Ian. 1981. "Tuvan Reindeer Husbandry in the Early 20th Century." *Polar Record* 20, no. 127: 337–351.
Wiklund, K. B. 1918. "Om Renskotselns Uppkomst." *Ymer*, no. 3: 249–73.
Ziegler, Jon Henrik. 2018. "Pigs and Spirits in Ifugao: A Cosmological Decentering of Domestication." In *Domestication Gone Wild: Politics and Practices of Multispecies Relations*, edited by Heather Anne Swanson, Marianne Elisabeth Lien, and Gro B. Ween, 50–71. Durham: Duke University Press.
Zolotarev, Aleksandr M., and Maksim G. Levin. 1940. "K Voprosu o Drevnosti i Proizkhozhdenii Olenevodstva [Concerning the Antiquity and Origin of Reindeer Breeding]." *Problemy Proiskhozhdenia, Evoliutsii i Porodoobrazovania Domasnykh Zhivotnykh*, no. 1: 171–189.

Part I

TOZHUS, TOZHU DISTRICT, REPUBLIC OF TYVA

Chapter 1

The Wolf-Pack Approach

Tozhu Hunting Strategies in the Siberian Taiga

Tayana Mongush-Arakchaa

Anthropological studies have typically presented reindeer as a key domesticated species for Siberian peoples. For Tozhu reindeer herder-hunters, however, such a perspective ignores the important roles of dogs and horses. These species are equally vital and interdependent partners of daily life in the mountainous areas where Tozhus live. Many Soviet ethnographic descriptions of traditional organization in taiga nomadic societies were constructed around research focusing on reindeer, and they were dismissive of the importance of dogs and horses. Although most research stressed the important role of dogs in the autumn hunt, it often overlooked the nuances of dog-human relations. Several ethnographic studies on taiga reindeer herder-hunters are good sources on hunting dogs in the South Siberian taiga zone, particularly the works by Petri (1928b, 1930), Brandišauskas (2017), Tugolukov (1969), and Oehler (2016). Vladilen Tugolukov (1969) gives an interesting account of the Laika (Indigenous dog breed), used among the Evenki of Buriatia. Bernard E. Petri (1928a, 1928b, 1930) describes hunting dogs in detail, including their training and the hunting process among Tofa in Tofalaria and the Tutur Evenki in Irkutsk Oblast'. Recent studies clearly demonstrate that among groups who are principally dependent on subsistence in the taiga, and for whom dogs are essential hunting partners, there exists some commonality in the nature of human-animal relations (Davydov and Simonova 2008, Oehler 2016). The tendency to focus on reindeer related research obscures our understanding of the complexities of the relationships between humans and nonhumans in the taiga environment, and it fails to give a complete picture of taiga societies. However, some anthropological studies of nomadic pastoral societies demonstrate a shift in theoretical orientation, which recognizes the need for an ethnography addressing interactions between humans and animals within particular socio-cultural structures (Fijn 2015, Nadasdy 2007, Stépanoff 2012). This multispecies sociality offers new insights into an understudied aspect of contemporary Tozhu lifeways and it adds nuance to current understandings of Tozhu hunting and reindeer herding strategies, challenges, and successes. This insight has become important to my understanding of the relationships between humans and nonhumans, which have become the primary subject of this work. From the beginning, this research was spurred by my interest in dogs as human companions. I soon realized that the classic anthropological question of the social organization of hunting and reindeer herding economies could be analyzed also from the perspective of human-dog relationships.

Tozhu reindeer herder-hunters, a small community of people in the Tozhu district of the Republic of Tyva in South Siberia live in partnership with multiple animals in the taiga zone. Tozhus speak the Tozhun dialect of the Tyvan language, which belongs to the Turkic language family. It was in fact the Tyvan language which gave the word *taiga* to the world. Northern reindeer herding is officially recognized as a leading industry in the Tozhu district, although the district probably has more cattle and horse breeders than reindeer herder-hunters. According to the data provided by the district administration, on January 1, 2018, thirty five people were engaged as reindeer herders, with a total of 2,667 reindeer (Krasnoiarskstat 2019). However, the district administration did not make a real effort to count the people and animals, and therefore the totals are very likely inaccurate. Typically, the count is done from what reindeer herder-hunters report. In reality, approximately 60 percent of the people on the official list are actively engaged in reindeer herding, and the remaining 40 percent are referred to by the actual herders as "paper reindeer herders." These people living in Adyr-Kezhig village own the animals but do not participate in reindeer herding. Typically they are kin of the actual herders, siblings, sisters, and brothers. Today, herder-hunters purposely keep forty to sixty reindeer because this number is sufficient to provide a household with transportation, milk, and other dairy products while still remaining manageable.

To better understand contemporary Tozhu herder-hunters' lives, I spent several months living in reindeer herding camps and in the village of Adyr-Kezhig. This firsthand experience provided me with knowledge of the Tozhu world, including their daily lives as well as a chance to observe their co-existence with reindeer, dogs, and horses. I immediately noticed that dogs play an important role in hunting, a part of history for Tozhu that remains largely unknown to the public. I began pondering when early taiga hunter-gatherers possibly first domesticated dogs and what dogs would have meant to them. This also raised questions about dogs in service to contemporary reindeer herder-hunters, and whether the use of dogs might be related to reindeer domestication.

This chapter focuses on the hunting dogs, commonly used by taiga reindeer herder-hunters. Taiga reindeer herding is quite different from tundra reindeer herding. Taiga reindeer herding is based on small-scale reindeer herds and reindeer are used as transportation while the other practice uses large-scale herds and focuses on meat production. The Tozhu herder-hunters have a deep relationship with reindeer, as well as with dogs. This chapter seeks to illustrate not only human-dog relationships, but also the synergistic effectiveness of dogs working alongside the herder-hunters, the success rates of hunting with dogs, and the strategies herder-hunters employ using hunting dogs. In order to advance our understanding of these topics, it is necessary to study both human and animal behavior in the taiga reindeer herding and hunting system.

DOMESTIC AND WILD

Tozhus, living in close relationship with a variety of animal species, offer us a unique view on human-animal relations and how humans can configure animal presence in the taiga environment. Tozhu hunting and reindeer herding practices have adapted to reindeer, horses, and dogs in order to fit the needs of humans and animals. Moreover, herder-hunters and their animals are bound in mutual relations, in which humans and animals have reciprocally influenced each other. As Keith Dobney and Greger Larson (2006, 261) stress: "Often in domestication studies, the terms wild and domestic have been interpreted as complementaries and not as antonyms, despite the fact that 'wild' and 'domestic' represent the extremes of a process and not a simple dichotomy."

Tozhu distinguish wild from domestic animals, while they recognize that their domestic animals cannot easily be understood as strictly domestic or wild. Although these animals (reindeer, dogs, and horses) are integrated into the human sphere and labeled as domestic animals, they were never removed from their original habitat. For instance, reindeer and horses are allowed to roam freely in pastures, and they return to the camps all on their own. Dogs also are often allowed to wander freely in the taiga where they hunt wild animals, returning to their masters on their own. In short, Tozhu have developed a flexible balance for properly managing each animal in both captivity and freedom.

The categories of "captivity" and "freedom" are central to our understanding of Tozhu hunting and reindeer herding practices. When it comes to hunting and herding, the other basic social concepts of "trust" and "domination," which are connected to "captivity" and "freedom," have become prominent in scholarship on human-animal relations. Although scholars working on the topic of "trust-domination" share the same core set of questions, these two categories are subjects of ongoing, invigorating debates. Tim Ingold (1994) explores "trust" and "domination" among hunter-gatherers and pastoralists, placing them into categories of two types of engagement with animals. Ingold suggests that the relations between hunters and wild animals are reciprocal relations based on trust. Thus, "a hunt that is successfully consummated with a kill is taken as proof of amicable relations between the hunter and the animal that has willingly allowed itself to be taken" (Ingold 1994, 69). In contrast, human-animal relations in pastoralist cultures are founded on the principle of domination.

Ingold's work is a stimulus to anthropological studies of those communities dependent for their subsistence on domestic or wild animals. His work also provides room to explore further theoretical implications. Some scholars challenge his theoretical framework, suggesting a need to rethink such basic concepts as "exchange," "personhood," "trust," and "domination" in the context of human-animal relations. These scholars point out the existence of a contradictory position of hunting in varied Indigenous cultures, and that, in particular, concepts of "reciprocity," "trust," and "domination" can contradict each other (Brightman 1993, Knight 2012, Nadasdy 2007, Oma 2010). In his study of human-animal sociality in an Indigenous North American context, Paul Nadasdy (2007, 27) describes a tension between negative dominance and positive reciprocity: "Although hunters do indeed sometimes see themselves as engaged in a relationship of reciprocal exchange with their quarry, at other times these same hunters express beliefs and engage in practices that seem to deny such a relationship."

The idea that animals give themselves to hunters as a type of reciprocal exchange is often noted in anthropological literature. The main part of this fascination is a romantic stereotype, viewing almost all hunting societies with existing human-animal relationships as practicing a form of metaphorical, reciprocal exchange.

Kristin Oma, exploring Scandinavian Bronze Age case studies, suggests an alternative model of how humans and animals have related to and interacted with each other. She argues that, "the development from hunting to husbandry signified a shift from domination to trust" (Armstrong Oma 2010, 177). Thus, trust is a missing element during the hunt of wild animals: "Although the hunter might trust the animal, from the animal's viewpoint this trust is not reciprocated; the prey flees in the presence of humans and exhibits no trust" (Armstrong Oma 2010, 177). Alternatively, in pastoralist societies, human-animal relations are based on trust: "The intimacy that is formed in the proximity between humans and domestic animals is highly dependent upon trust.

Humans trust animals to be docile and cooperative, while animals trust humans to protect them, feed them and care for them" (Armstrong Oma 2010, 177).

John Knight (2012), drawing on biologist Robert Hinde's model of relations, argues that hunting provides minimal opportunities for developing a personal form of relationship; these are usually developed within a human band or group. Because of the nature of the hunt, a hunted animal avoids encounters with the hunter: "Encountering one another for the first time, hunter and hunted are literally strangers. What the hunt, as an event, simply does not allow is for the hunter to build up a 'file' on the individual animal he happens to come across" (Knight 2012, 340).

In the hunting setting, human-animal relations mean that the hunter gets to know prey animals only as a class of animals, and builds up general knowledge about animal behavior, as opposed to specific knowledge of any particular animal. The hunter's interests fundamentally conflict with the interests of the prey animal.

In the case of animals with which herder-hunters interact in the taiga, both principles, "trust" and "domination," can be observed, though the widespread idea that animals give themselves to humans is not shared among Tozhus. During my fieldwork I expected to hear stories describing rituals or traditions for obtaining luck before the start of the hunting season, making offerings to taiga spirits (Tyv. *taiga eeziler*),[1] and conducting sanctification ceremonies performed by herder-hunters or shamans. To my disappointment, I did not hear such stories except for a brief description of a sanctification ceremony held by a Tozhu shamaness several years ago. When I asked my host, Sayzana,[2] what they thought about taiga *eeziler* and how they were to be treated, she told me that her family was too busy with hunting to think about them. Sayzana also stressed that if she and her kin would think about taiga *eeziler* all the time, they would not be able to hunt or live in the taiga. Although Tozhus live in the taiga, they recognize it as a foreign and quite adversarial place for humans to live. In short, these hunters think of themselves as intruders, rather than residents of the taiga realm. The Tozhu relationship with spirits reflects a stratified society ruled by numerous powerful *eeziler*, in which humans have an inferior social status.

Social stratification characterized Tyvan society before it joined the USSR in 1944. This social stratification is also reflected in the Tozhu understanding of human-nonhuman relations. According to Svetlana Biche-ool (2007), before Tyva became a Russian protectorate in 1914, Tyvan society was composed of several social classes. The most important of these classes was the political economic elite, which was divided into segments: *noyons,*[3] including *lamas,*[4] and officials. The governing class was composed of the *khozhuun noyons*;[5] each surrounded by an administrative apparatus. The official political leader of the Tyvan society was the *ambyn-noyon*. This elite class accumulated wealth by imposing taxation on ordinary nomads and exploiting their labor. *Noyons* controlled 8 percent of the best pasturelands, owned two-thirds of livestock, and 60 percent of the sheep and goats in Tyva (Biche-ool 2007). This hierarchical structure applies to the *eeziler*, and their hierarchies of the spirit world. The two ruling classes, *noyons* and *eeziler,* hold equivalent positions at the top of the hierarchy, and the taiga nonhuman society shares structural features with the human nomadic hierarchy. Much like the ruling class of *noyons*, *eeziler* have an elaborate status in the culture distinguishing them from other nonhuman and human members of the taiga society; they own wild animals and control all the wealth of the land. Each *eezi* owns a particular natural object: a lake, a pass, a river, a mountain, a valley, a pasture, a forest. Taiga *eeziler* are considered the most powerful. In fact, the higher the mountain, the more powerful its *eezi.*

How does the dominant nature of *eeziler* and resource-seeking behavior of Tozhu herder-hunters co-exist in the taiga environment? If humans truly have lower social status in the taiga, how do *eeziler* interact with the hunting Tozhu? Although *eeziler* are powerful and strong beings, they are not in a real position to compel herder-hunters to work or create wealth for them. Tozhu hunting success depends on cooperation and sharing behaviors derived from a strategy of using domestic animals. Together with their reindeer and dogs, herder-hunters enhance their hunting opportunities, and also expect *eeziler* to share resources. Reindeer, dogs, and horses have special status; they are domestic animals, but they are considered to be half-wild animals. Acknowledging interdependent animal autonomy, Tozhu try to maintain this feature by properly balancing between their captivity and freedom. Additionally, captivity fluctuates seasonally depending on household needs.

It is important to stress that Tozhu life has dramatically changed over the past eighty years, largely in response to sedentarization and collectivization imposed by the Soviet government. Many old practices regarding life in traditional dwellings, performing rituals, and maintaining food preferences have fallen into disuse. Today Tozhu herder-hunters reside in sedentary villages where their families live in houses. Tozhu living in villages can rely on their extended kin nearby to look after and protect their interests. However, these recent changes have had relatively little impact on ideas about human-animal relations. The Tozhu see a distinct separation between the sedentary villages and the taiga camps, and taiga mobile activities have greater prestige than urban activities in the village. In taiga settings, although herder-hunters are often neglected by *eeziler*, humans still seek to make allies with nonhumans. People often count on the support of *eeziler* in hunting activities, although these supernatural beings are quite capricious. At the same time, taiga dwellers often fear all contact with *eeziler.* While reading ethnographic Tyvan literature, I noticed a number of stories or epics that described encounters with supernatural beings that had happy endings, particularly when *eeziler* give many wild animals to lucky hunters. However, from my own experience, hunters in the Biy-Khem district (a neighboring district to the Tozhu district with whom Tozhu share similar hunting traditions) always expressed fear of meeting *eezi* or being punished by them. While I lived in and visited the Biy-Khem district (also in a taiga zone) I heard only stories about *eeziler* inflicting punishments including suffering and death. In addition, I heard a story of one female Tozhu punished by a taiga *eezi.*

Both Tozhus and Tyvans in Biy-Khem often think that contact with an *eezi* can bring punishment.[6] This human-*eeziler* avoidance is associated with pressure on taiga resources, particularly pressure due to intense hunting. Tozhus believe that the animals never give themselves willingly to human beings. Moreover, *eeziler* often protect their animals from human hunters. In addition, herder-hunters realize they are part of the food chain; which is why they never leave their camps without a firearm, even if they stay in the near vicinity. What should hunters do to overcome the power of supernatural beings in order to capture wild animals? The solution Tozhu have found is to extend their hunting techniques by creating alliances with other sentient beings. Tozhu are not just hunters; they are hunters working in cooperation with reindeer, dogs, and horses. They intrude into the taiga with these animals, and only with their aid can Tozhu herder-hunters act as super predators, equivalent to apex predators such as wolves.

In the Tozhu case, a successful hunt involves not only the obtaining of wildlife but also alliances with reliable hunting partners, humans, and animals. For Tozhus, who have spent all their lives in the saddle, reindeer symbolize the Tozhu nation, and control of reindeer, dogs, and horses parallels control of the taiga. A herder-hunter who cannot master reindeer, dogs, or horses certainly cannot master unruly wild animals, let alone his own family or clan. Thus, alliances with multiple animal species, in the context of domestication, can be understood as fluctuations of dominance and trust through captivity and freedom. For Tozhus, dominance and trust are necessary means in the taming process and cannot be distinguished from domestication in general. The ancestors of the Tozhu knew that reindeer were not born to carry loads or human passengers, and that these animals would be stubborn in their resistance to performing these tasks. But those archaic hunter-foragers learned that they could develop a docile reindeer that could be easily managed. Bradshaw (2011) points out that "domestication has not adapted dogs to human environments; it has merely given them the means to adapt" (Bradshaw 2011, 122). Although this statement refers to dogs, it can be applied to other domestic animals as well. Herder-hunters learned to give their reindeer the means to adapt to the human environment; they improved the practicality of the breed and bred reindeer that suited their needs. In fact, today, taiga reindeer cannot be easily categorized as either domestic or wild. They are not held in full captivity, and they exist in their natural environment with very little human monitoring. In the Tozhu case, the entire taming process is based on keeping reindeer in partial captivity; animals are not locked in the pens and can roam freely across pasturelands. In other words, herder-hunters developed the right balance between a captive, supervised structure and letting the animals roam freely in the vicinity of their camps. Typically, captivity is viewed by scholars as a state in which animals are kept within human-made barriers. The taiga type of captivity is different; it is straightforward and simple—animals are tethered to stakes or tree trunks for long hours. In this case, the reindeer is kept in captivity in the same way as if it were kept behind a fence. This taming process starts two to three days after calves are born and continues for their entire life span. The Tozhu perceive this tethering of calves and reindeer as an essential practice for cultivating that most valued attribute—tameness. What does tameness mean in reference to reindeer? In general, it means that the herd is easily managed and taught. From an early age, calves learn gradually how to cope with and adapt to a human environment. The timing of the first introduction to tethering is absolutely crucial to how reindeer react to handling later on, particularly by staying close to the *aal*, carrying loads and humans, and tolerating milking. After a month, the calves no longer resist being tied up.

To carry humans and loads of supplies, the reindeer must be calm, patient, docile, and easily handled. For thousands of years, early herder-hunters selected bulls for tameness; today modern herder-hunters continue to follow this criterion scrupulously. They do not allow rebellious bulls to breed; such bulls are usually the first choice for slaughtering. Selective culling is especially important in maintaining the tame nature of the reindeer. As with most domestic animals, morphological changes accompany domestication; this can be seen in the basic morphological, physiological, and behavioral changes reindeer have gone through over time. Reindeer bodies have altered; they have become shorter and of stouter stature than their wild long-legged relatives. The rhythm of the breeding season begins earlier for domesticated reindeer, by up to four weeks (Baskin 2009). Under such selective pressure favoring tameness, domesticated reindeer have inevitably become tamer then their wild counterparts. In general, Tozhus might be said to transgress the borders of wild and domestic; balancing these two categories by allowing animals to assist them in the best possible way.

Dominique Lestel et al. (2006) advocate combining ethology and ethnography in order to develop an account of the complexity of relationships between humans and nonhumans. They examine two disciplines in studying human-animal relationships in hybrid communities in which humans and animals live together in a shared environment: "Ethno-ethology recontextualizes the approach to modes of knowledge within the interactivity of human/nonhuman relations in order to identify the representations and other cultural phenomena humans use to interact with animals and the practices concomitant with these representations" (Lestel et al. 2006, 173). This approach also "seeks to describe and understand how humans and animals live together in hybrid communities sharing meaning, interests and affects, articulated around jointly negotiated significations" (Lestel et al. 2006, 173). The authors stress the importance of understanding not only how humans use animals but also "how humans and animals live together in that society—the way they use each other is only one dimension among others—and what kind of sociability results from this living together" (Lestel et al. 2006, 159). Lestel et al. interpret this "living together" as "shared complexity in interspecific associations," with understanding that "shared complexity" is "the phenomenon that characterizes interactions between natural or artificial and human or nonhuman organisms interacting with each other" (Lestel et al. 2006, 160). These "hybrid associations" represent communities in which humans and animals have shared meaning and interests and everyone is involved in social interactions (Lestel et al. 2006).

My approach to these human-animal relationships is similar to that of Lestel et al. (2006) and Natasha Fijn (2011), in that I consider herder-hunters and their animals as a "hybrid community" in which "agents who participate in a reciprocal, co-domestic relationship" share meanings, interests, and affects (Fijn 2011, 36). The ethnography of Tozhus demonstrates that human activities have influenced animal behavior just as animals have influenced humans. Cohabitation of humans and multiple animals has led to a human-animal mutual reciprocal relationship.

The works discussed in this chapter, particularly those by anthropologists, explore the processes of domestication and the patterns of animal and human behaviors in conjunction with each other, by examining how humans and animals adapt to living in shared environments. These studies focus primarily on non-Western societies and their domesticated species and evaluate the degree to which human-animal relations can be explained as adaptive responses to the habitats in which they live. A core question that emerged from the analysis of ethnographic data compiled during my research is the following: How are the relations of Tozhu herder-hunters with their reindeer and dogs best understood in terms of hunting and herding practices? To answer this question, I examined the practices and representations of Tozhu herder-hunters who follow what I call the "wolf-pack approach." Focusing on this approach has provided a fruitful conceptual starting point for understanding the larger framework in which Tozhus have learned to use the desirable features of domesticated animals, and more broadly for interpreting the co-existence of humans and animals in the taiga.

The wolf-pack approach suggests that herder-hunters mimic wolf hunting behavior while relying on dogs' abilities to track and capture their prey. Tozhus recognize dogs as being inherent hunters and pack members who must be controlled by a human acting as the ultimate pack leader, guiding the behavior of the dogs. Previous literature on dog behavior has incited controversy among scientists and dog trainers. This controversy has already raged for more than a century. While traditionalist trainers promote the wolf-pack approach, other cynologists are highly critical of the manner in which academics employ it to shed light on human-dog relationships. Famous dog behaviorist Cesar Millan and his co-author Jo Peltier (2014) state that it is crucial to understand a dog's pack mentality in interactions with other dogs and humans:

> A dog's pack is his life force. The pack instinct is his primal instinct. His status in the pack is his self, his identity. The pack is all important to a dog because if anything threatens the pack's harmony, it threatens each individual dog's harmony. If something threatens the pack's survival, it threatens the very survival of every dog in it. The need to keep the pack stable and running smoothly is a powerful and motivating force in every dog—even in a pampered poodle that has never met another dog or left the confines of your backyard. Why? It's deeply ingrained in his brain. Evolution and Mother Nature took care of that. (Millan and Peltier 2014, 76)

According to Bradshaw, the idea "that dogs, because of their ingrained 'pack mentality,' can be controlled only if their owners adopt the role of pack leader, and that the most reliable way to insure this outcome is by the use of physical punishment" (Bradshaw 2011, 96) has been challenged by recent canine studies, and it has been rejected by many scientists and dog trainers alike. My opinion is that in the specific context of the taiga environment, the wolf-pack approach remains fruitful and inspiring for the understanding of herder-hunters' relationships with their dogs, and that it can be extended to include other species living alongside humans, such as reindeer and horses.

We must keep in mind that our knowledge of animal behavior can be key to our interpretations of human behavior (Laland and Brown 2002). Lacking a wolf's fast legs, humans "borrowed" reindeer legs by learning to ride this animal. As a result, their two bodies blended into one, and the human became more of a "wolf" managing the hunt singularly. Therefore, the pack hunting pattern forced early hunters to domesticate reindeer, allowing them to form a human-led "wolf" pack. Today, Tozhu herder-hunters still strongly encourage this wolf-pack mentality in their camps.

"WOLVES" WITH REINDEER AND DOGS

There is more to survival behaviors than simply hunting efficiently, as a multitude of skills enable hunter-herders to thrive in the taiga environment. Understanding the methods hunter-gatherers use in hunting is central to measuring their hunting effectiveness, as well as their ability to survive and thrive. Killing animals for their meat and fur with the aid of reindeer and dogs was an important behavioral modification of early hunter-gatherers. Stépanoff and Vigne (2018, 4) point out that, "In the various centers of domestication throughout the world, people know of narratives explaining the close relationship formed with certain species." According to Darzha (2013), the early population in the territory of Tyva developed a successful hunting system by mimicking the hunting style of other apex predators—wolves. In Inner Asia, Turkic tribes consider the wolf so symbolically central that it is sometimes referred to as their totem. For instance, Mongolian tribes believe that their lineage begins with the pairing of a wolf and a deer (Kahn and Cleaves 1998). The Turkic nomadic tribes had an ancient legend of the wolf and a boy:

> The story told of a female wolf finding a human baby boy whose feet had been cut off and who had been abandoned on the steppe to die. The mother wolf nursed the boy back to health, protected him, and reared him. When the boy grew older, there was no one else to love him, so he mated with the wolf. From their offspring descended all the Turkic tribes that spread out from Mongolia. From them arose all the notable Turkic nations of history. (Weatherford 2010, 244)

According to Benedikte Møller Kristensen (2015), Dukhas believe that their clan originated from a wolf, which is why they are not allowed to kill wolves. This lineage history can be interpreted as suggesting that the first domesticated animal was a dog (wolf-dog) from the Saian-Altai region. This mythic kinship depicts closely connected relationships between humans, dogs, and deer. The Saian-Altai peoples believed that wolves were the first survival instructors of humans. People valued wolves highly as efficient hunters and intelligent teachers. A wolf is referred to in Tyvan as *böry̆*, *kokay*, or *kök-kara* (Tyv. with blue eyes). In Tyvan belief, wolves were the first hunting instructors. In order to perform extremely difficult survival tasks with less effort, early humans mimicked and modeled their hunting methods after the swift pursuit of wolves. Shipman (2015) argues that climate change and the arrival of modern humans, who were "an invasive new apex predator" (Shipman 2015, 230) with the ability to form alliances with wolf-dogs, caused the extinction of Neanderthals. It appears that Neanderthals "were slow to innovate and slow to change their ways in their world"

(Shipman 2015, 229) and did not form alliances with other species as a survival strategy. As Shipman notes,

> Although modern humans belong in the order Primates, not the Carnivora, and do not possess the slicing or crushing teeth, strong jaws, powerful limbs, and sensory abilities that most carnivores do, behaviorally Homo sapiens is unquestionably predatory. Our ability as predators hinges on our ability to make and use tools to perform the activities for which carnivores use their jaws, senses, legs, teeth, and claws. (Shipman 2015, 72)

According to Shipman (2015), one of the most striking behavioral and technological breakthroughs of modern humans emerged from modeling wolves' hunting style. Once again, this part of the story of early taiga hunter-gatherers' survival begins with a remarkable achievement—the domestication of canids and ungulates, and the development of cherished multispecies relationships between humans and nonhuman creatures. Typically, humans expanded their abilities by gaining the good sensory abilities of canids, but in taiga conditions they needed to use the swift strong legs of ungulates as well. Modern humans succeeded in domesticating the carnivore and would thus have been pre-adapted to domesticating ungulates to gain assistance in hunting. Later, hunter-foragers of the taiga zone succeeded in building quadrangular relationships between humans, dogs, reindeer, and horses. They would have accumulated knowledge of these animals' behavior to improve their own hunting skills.

Scholarly findings suggest that wolves could be domesticated because of their ability to cooperate and communicate with their human masters (Hare et al. 2002, Miklósi and Topál 2013). The main impact of wolves on human life came through humans modeling their hunting tactics. The Tozhu hunting process with dogs is analogous to a wolf-pack approach, in which, in order to achieve hunting success, herder-hunters imitate the behavior of wolves. Wolves hunt cooperatively by chance or by smell, particularly preying on large ungulates such as deer, moose, and elk. For instance, when hunting a roe deer, wolves locate their prey, chase it, surround it, and kill it (Heptner and Naumov 1998). Humans rely on similar collective tactics to capture large animals. In addition, in Inner Asia they cooperate with different hunting partners—dogs, horses, eagles,[7] and reindeer. I propose that imitation of wolves' hunting tactics was a stimulus to domesticate not only wolves but also ungulates and raptors. The greatest talent of Eurasian nomads lay in their ability to use the predatory strengths of other species—eagles and dogs—to their collective advantage. Horse-riding and eagle falconry are some of the highest achievements in the hunting history of humankind. Darzha (2013) states that every man can hunt, but only a few can hunt while simultaneously coordinating a horse, an eagle, and dogs. This type of hunting requires enormous technical skill and the cultivation of a hybrid knowledge of three separate animal species. Using all of these species, hunters capture prey faster and with less effort during their hunting trips. For example, horses are the fastest means of transportation in the steppe, and eagles play an important role in quickly locating prey, such as hares, foxes, and wolves. Dogs are used to find the prey and keep the target animal in one place by surrounding it and barking until hunters approach and shoot it. Moreover, in the past hunters used two breeds of hunting dogs simultaneously. The *Taigan* and the *Kadarchy Yt* each perform different tasks in the hunt for wolves. Light *Taigan,* as pointers, chased down anything that moved, and large *Kadarchy Yt* were specifically used for big game hunting (Darzha 2013). As a result, this type of hunting became more complex. Humans train dogs and eagles to cooperate with each other. Dogs coordinate with eagles in locating the animal, and eagles are trained to hunt wolves and foxes. Looking at this hunting practice, we can see a continuum of multispecies relationships of active hunting partners: dogs, eagles, horses, and prey animals, the latter of which "are the ultimate objects of hunting but are nonetheless capable of understanding human behavior, evading capture, or fighting back" (Simon 2015, 695).

Although hunts requiring intelligence and collaboration with four nonhuman species were common in Inner Asia, eagles are largely useless in the taiga because they cannot locate prey in deep forest. Dogs and reindeer played a significant role in adaptive hunting strategies in the taiga zone. This is the key transformation of the taiga zone: the evolution of the "hunting triad of dog-reindeer-human" (Anderson et al. 2017, 8), committed to the killing of wild animals. If we consider a herder-hunter a "wolf" lacking strong jaws, crushing teeth, fast legs, sharp vision, good hearing and scent, this "wolf" has to cooperate with the other pack members—in this case dogs—to take down large prey which may otherwise flee. As in the wolf-pack, the human "wolf" as leader defines the role for each pack member and together they work as a team. With minimal communication they understand each other and collaborate so effectively that the pack becomes a high performing team. Tracing the scent and trail of an animal, the dog is located at the front of the triad (Anderson et al. 2017).

In actuality, the human leads the dogs to the kill site and subsequently trusts them to lead the triad in the hunt (see figure 1.1). The "wolf" releases the dogs in pursuit of the targeted quarry. Their task is to locate the target animal, chase it over long distances, surround it, and give a signal to the "wolves" by barking. Reindeer are also active hunting companions, providing a reliable means of transportation to move swiftly through the taiga. In addition, some older, experienced reindeer, hearing the dogs' barking, can locate them and bring the "wolf" to the place where the prey is surrounded (humans

Figure 1.1 "Hunting from the Viewpoint of a Fellow Carnivore." *Source*: 2019, A. Arakchaa.

do not possess the acute hearing of the reindeer) (Tugolukov 1969). While humans and nonhuman animals can mimic each other, humans lead the main mimicking role in the hunting activity. The Tozhu imitation process is the inverse of Rane Willerslev's description of the Yukagir hunters: "By means of mimicry, the Yukagir hunter assumes the viewpoint, senses, and sensibilities of his prey while still remaining aware of himself as a human hunter with the intention of killing it" (Willerslev 2007, 26). In the Tozhu case, the herder-hunter assumes not only the viewpoint of his prey but also assumes the viewpoint of a fellow carnivore, particularly by thinking about each pack member's hunting behavior, considering age, experience, gender, and social status in the "pack," as well as the ways in which each can contribute to the chase and kill. Thus, the herder-hunter must keep in mind many factors to achieve hunting success and to avoid wasting energy.

Herder-hunters working with reindeer and dogs capture not only more prey, but also a wider variety of prey. They have an enhanced ability to take down large prey and bring it home from long-distance hunting expeditions. Cooperation with these animal species helps decrease the human expenditures of energy in hunting. Not only do humans and dogs cooperate with each other, but dogs and reindeer do as well. Hunting without reindeer and dogs must have required enormous strength and skill in taiga conditions; it can also be dangerous with some species, such as boar and bear.

If early hunter-foragers spent thousands of years affecting reindeer breeds, then the reindeer would also have changed the hunters. Evidence indicates that taiga hunter-foragers tend to be smaller than their neighboring cattle breeders. Petri (1927) noted that Tofa were of a lighter weight, and it was seldom that a Tofa man's weight would reach about four *poods*[8] (65 kilograms). The weight of a rider is crucial, especially when a hunter spends long hours searching for prey in the mountainous and hilly taiga. Carrying a heavy rider, a reindeer will soon become exhausted. In fact, there is always the risk that a heavy man could break a reindeer's back. Although forest people are smaller around the world, I suggest that the reduction of body weight was an adaptive strategy, which allowed Tozhu to live together with the reindeer, and specifically to ride them.

Views regarding stature and mating height preferences in many human populations are well documented by other researchers (Courtiol et al. 2010a; 2010b, Jackson and Ervin 1992, Pearson 2000, Salska et al. 2008). Alexandre Courtiol et al. (2010a) note that height and stature play an important role in the perception of attractiveness in human mate choice. Height and body mass significantly influence perceptions of an individual's body shape. Moreover, the effect of body mass on body shape varies between women and men. Courtiol et al. (2010b) chose height to determine variations in mate preference among French men and women through selection experiments. The findings indicate that both men and women prefer mates who are significantly taller than average. Jackson and Ervin (1992) examined the height stereotypes among women and men in the United States. Their study results reveal that both women and men consider shortness a liability. Irmina Salska et al. (2008) examined height preferences in 2,000 personal ads in the United States. Their study demonstrates that height preferences are strongly influenced by one's own height as well as by societal norms. The researchers stress that society has a set of expected gender norms about physical appearance based on the assigned sex. Thus, women generally expressed a preference for the taller-male norm because tall stature provided advantages to offspring. Men also preferred a shorter-than-male female partner, as men were considered more powerful than women (Salska et al. 2008). Although these studies were completed among sedentary populations living in industrial societies and having a vastly different lifestyle, similar selection criteria can be seen in Tozhu society where men are also expected to be taller than women.

Tozhu hunter-gatherers need to be in good physical shape to perform such tasks as reindeer riding and hunting and walking in mountainous and rugged taiga-tundra. While there are tall Tozhu men, the majority are short. A herder-hunter's light weight always coincides with a physically strenuous lifestyle, engaged in subsistence activities. With a relatively light weight a herder-hunter still must be able to control a reindeer that weighs 200 kilograms and moves swiftly in hilly taiga. A lightweight rider helps to decrease the load on a reindeer. Most Tozhus tend to be fairly short, and their stature seems to be due their occupational weight limits.

During my research I documented these cultural and physical aspects of reindeer riding. Herder-hunters would ask about my weight. It is certainly rare in fieldwork that a researcher's weight is so critical. When I first came to Sayzana's *aal*,[9] Anoka, Sayzana's brother, asked what my weight was, and I said that I weighed 64 kilograms (my height is 161 centimeters). He told me that my weight was fine, and a reindeer castrate (Tod. *chary*) by the name of Pensioner would be able to carry me. According to herder-hunters, an ideal rider's weight should not exceed 60 kilograms, although a large reindeer[10] can carry a 70 kilogram rider. For both horse and rider safety, the Certified Horsemanship Association recommends the 20 percent rule, which means the rider's weight should not exceed 20 percent of the horse's weight. For example, a 400 kilogram horse can carry a 90 kilogram rider (Certified Horsemanship Association 2017). A trained riding reindeer weighing 200 kilograms can safely carry a rider of 60 kilograms, which is 30 percent of the reindeer's weight. Therefore, the Tozhu seem to use a 30 percent rule. If a reindeer is large and strong and can carry a 70 kilogram rider, the reindeer carries 35 percent of its weight. In addition, in winter, the riders have heavier winter clothing, which adds extra weight to the reindeer. Although the social acceptance of burdening an animal differs between Tozhus and the Certified Horsemanship Association, the rules are comparable. If we compare the 20 percent horse rule with the 30 percent reindeer rule, it would seem that the reindeer are stronger.

One day Vika, Anoka's wife, noticing that I liked to eat candy, commented: "*Kaanbet chiir bolgash semis-tyr sen*" (You eat candy; that is why you are fat). It was not flattering to hear that I was fat. However, I felt a little chubby around such fit people. In addition, the Tozhu are noticeably smaller than Tyva living in other parts of the republic. I felt that almost everyone was a little bit shorter than I, and only met two men who were taller, Andrey and Cheyneshko (a man from the neighboring camp). Both were slender and thin boned. Because I did not want anybody to question my weight on the next field trip, I purposely lost 4 kilograms so as not to tax the reindeer.

When people asked about my weight, I was always reminded of a story that took place with my ex-husband, Orlan, who at the time was working as a policeman in Turan, in the Biy-Khem district, where we lived between 1995 and 1996. During that year, Orlan often had to chase down *kaygals* (Tyv. for cattle thieves, rustlers) on horseback in the taiga. It was a challenging and quite dangerous task because some *kaygals* were treacherous and would shoot chasing herders or policemen from an ambush. In the fall of 1995, several cows were stolen from a herd, and it was known that the rustlers were from the Tozhu district. Orlan and his two colleagues were sent to pursue the *kaygals* in the Tozhu district. The group penetrated deeply into the taiga, reaching a reindeer herding camp. There were two traditional dwellings (Tyv. *Chadyrlar*),[11] and the policemen asked the herder-hunters for permission to stay for the night. When it was time to go to bed, the hosts showed them a place to sleep in the *chadyr*, but they told Orlan, "There is no place for you here. A young woman with a toddler lives in another *chadyr*. Go to sleep there; she needs to have a son." Orlan was shocked by this straightforward statement; he did not know what to say and looked at his older colleagues. The men were silent, and Orlan had to go to the other *chadyr* thinking about his awkward predicament with these Tozhus who seemed so radically different from the urban population he knew. He was on a mission to catch rustlers, not to make a baby with a completely unknown woman. In a way, however, it may not have been surprising that these people chose Orlan as a father for the young woman's future child. At age twenty-two, he was handsome, healthy, slim, and athletic, and his height was 176 centimeters in contrast to his colleagues who were in their late thirties and early forties, less attractive and of greater weight. The young woman was in her mid-twenties. She showed Orlan a place to sleep on the right side of the dwelling, the male side, as she went to sleep on the left female side of the *chadyr*.

The next morning, an old woman approached Orlan and said: "I am so glad that you did not sleep with the girl. You are big-boned. If she would carry a son from you, he would have big bones and be heavy, too. When it would be time to ride a reindeer, he would break the reindeer's back. It would be disastrous for us to have such a man in the aal. They did not think properly when they sent you to her." Apparently, the people did not see, right away, that the young man had big bones. But what does it exactly mean to have big bones? It seems, the elder was trying to say that big-boned Orlan would pass on his big bones to a new child.

In general, a tall and big-boned man is heavier than a short, small-boned man. It would be a challenge for the family to have a heavy, big-boned man in a household in which the economy depended on hunting, and in which men on reindeer spent their days searching for prey (see figure 1.2). A rider's weight is a critical criterion for riding a reindeer and for finding a mate with whom to start a family, from whom to inherit one's genes. It is necessary, at this point, to mention that Tyvans recognize the role of a father as one who gives bones to a new child, whereas a mother contributes flesh and blood. For Tyvan groups, belonging to a tribe was very important. Since they were patrilineal societies, a male lineage or a clan was known as *söök or seok* (Tyv. for bone, bony, breed, quality) literally meaning "bone of the father." In other words, the origin of a child's bones signified affiliation with one's clan or tribe.[12]

Figure 1.2 Small-Boned Sergey with his Small Stature Riding a Reindeer. *Source*: T. Mongush-Arakchaa.

The aforementioned story demonstrates that Tozhu believed that the man created not only the bones but also the quality of the bones: either thick or thin. The Tozhu elders, recognizing the role of the father as the bone giver, preferred their daughters' partners have short height and small bones. What makes the end of Orlan's story interesting is that later we had a daughter who grew to be big-boned and wide-shouldered, although I am small-boned. The elder had foreseen that Orlan would pass his big bones to his child.

Weight limitations for riders are not only an important safety measure for reindeer. Small riders are also better able to maneuver an animal in difficult terrain. Tozhu generally perceive riding skills as key to being a skillful hunter. A successful human predator and pack leader maneuvers well and is able to capture prey while moving through challenging terrain. With their large reindeer and smaller human stature, Tozhu may be among the most well-adapted hunters living in the boreal forest.

In spite of generally avoiding their cattle and horse breeding neighbours, Tozhus are known to have engaged in trade with them (Bogdanova 1962). From this, intermarriages have emerged between different clans and tribes. From the story above, one might assume that in the past, elders would have observed the physical features of potential partners from other clans and tribes, particularly noting a suitor's weight, height, and bone density—all crucial factors in riding a reindeer. The prolonged co-existence of humans and reindeer required adaptation on both sides. This process was not exerting a one-way human influence on animals; hunter-foragers themselves adapted to better suit their reindeer's morphology. Tozhu not only cultivated docile deer, but having bred the world's largest specimen, they themselves evolved into smaller human riders. As a result of the long process of reproductive selection and the demands of their way of life, Tozhu generally exhibit smaller bones, shorter height, and lower body weight.

Both Michail G. Levin (1958) and Sevian I. Vainshtein (1961) noted that Tozhu differed in their appearance from the remainder of the Tyvan population. Napoleon Wolánski and Ewa Kasprzak (1976) point out that poor living conditions also contribute to the short height of the Tozhu population. In the taiga, filling the predatory role of "wolf hunter" requires that Tozhu find and pursue their animal prey; small stature and light body weight allows them to move more efficiently with their reindeer than taller people. The above related personal account suggests how a recognized preference could have contributed, over the long term, to changes in overall size as an adaptive strategy, with Tozhu elders articulating reproductive preferences between clans and tribes that would allow them to adapt to a life among reindeer in which the power of these animals could be harnessed to their advantage as hunters.

This research project attempted to move away from the dominant emphasis on reindeer in the daily practices of contemporary Tozhu herder-hunters. Such views have encouraged the propagation of an incomplete picture of the taiga dwellers and neglected the social complexities that define taiga human-animal communities. The idea that reindeer are the only one out of several important animals in taiga reindeer herding communities is grounded in the Soviet administrative discourse of northern tundra reindeer herding and early Soviet ethnographic accounts. Generally, these accounts represented reindeer as the only economically profitable animal that could be sustained in the taiga. Moreover, ethnographic studies about Tozhu have largely failed to depict other animals as interdependent partners in the daily life of herder-hunters and their mutual sociability resulting from living collectively. My research leads me to suggest that we should instead see Tozhu herder-hunters from a multispecies perspective, through a prism that balances the importance of reindeer with other important animals, most significantly dogs. I have attempted to do this by emphasizing what I have called the wolf-pack approach.

A number of researchers have proposed economic and ecological adaptation models, describing the domestication of reindeer based on taiga dwellers' imitation of their horse and cattle breeder neighbors to the south (Ingold 1980, Laufer 1917, Vainshtein 1961, 1980, Vasilevich and Levin 1951). We can see that they ignore one important ethnographic fact, which is easily revealed by a closer reading of the pastoralists' mythology and cosmology. These scholars miss the reality that reindeer domestication is connected more fully to an imitation of wolf-pack hunting behaviors than simply to horseback riding. The heart of this multi-species ethnography of Tozhu lies in the ramifications of mimicking different species' efficacious behaviors, particularly hunting strategies, and including such actions as the sounds they produce. I argue that, in the Tozhu case, we are presented with double and multiple mimicries, as the early hunter-foragers successfully mimicked wolves' hunting strategies, while simultaneously borrowing equine riding techniques from their livestock-raising neighbors, who practiced horse-riding eagle falconry techniques. Without the strong collaboration of dogs, reindeer domestication would have been less plausible. It is well-known that dogs were domesticated because of their hunting abilities (Clutton-Brock 1981, Gupta 2004), and hunting activities with these hunters became a stimulus for reindeer domestication; thus, reindeer domestication developed in support of hunting activities in the taiga. Such taiga "hybrid associations" represent an analog "wolf-pack" community, in which humans and animals are involved in mutually beneficial relationships of cooperation. Moreover, dogs, reindeer, and horses complement each other's abilities, depending on the season and human needs.

As previously mentioned, the wolf-pack approach came under attack from promoters of reward-based methodologies in the 1990s. However, it would be a misconception to see the Tozhu wolf-pack approach as being built only on expressions of human dominance. In a combined human-dog-reindeer pack, the human needs to be an effective team leader to achieve a successful hunt. In fact, Tozhu not only understand wolf-pack mentality but also apply it to their own hunting when working collaboratively with other pack members. This team approach includes elements of dominance, trust, love, and shared hunting passion. The job of the herder-hunter is to coordinate the hunt for each pack member. Dogs will locate prey, chase it down, and give a signal to the pack leader, who follows after them with his weapon. As a rule, the human leader kills the prey and rewards his canine members by sharing the prey with them. There is some degree of dominance and hierarchy in human-animal relationships, in which a herder-hunter will assert leadership, including physical punishment of a dog when necessary. Nevertheless, my data demonstrate that herder-hunters have largely benevolent relationships with their animals, based on trust, because dogs are perceived as independent entities and reliable hunters in their own right. Hunting with dogs is an expression of the trust that exists between pack members, the more a hunter and his dogs hunt, the more they get to know and respect each other. A herder-hunter has deep knowledge about each dog's hunting competency, prey-specific skills, experience, and personality.

The Soviet ethnographic descriptions of traditional organization in taiga nomadic societies were constructed through research of the reindeer, and were dismissive of the importance of dogs and horses. Although all research stressed the important role of dogs in the fall hunting, they often overlooked their crucial roles. The use of hunting dogs in the taiga zone remains an understudied topic in cultural anthropology. There are some ethnographic records of the use and treatment of hunting dogs in indigenous communities of southern Siberia which makes it possible to compare past schemata with the present in order to see the dynamics of change in hunting and reindeer economies. More research is needed to fully understand the role of hunting dogs in contemporary local economies and human-animal relationships.

NOTES

1. *Eeziler* is the plural form for *eezi* or "supernatural spirit" in Tyvan.
2. I stayed with Sayzana Kol and Andrey Baaran, a couple of reindeer herder-hunters.
3. A wealthy nomad, a ruler of a particular administrative unit.
4. A Buddhist priest.
5. District *noyon* or *ukherida*.
6. To obtain luck before the hunt, herder-hunters make offerings to the taiga spirits or perform rituals to please a taiga *eezi*. If the eezi is pleased he or she brings luck to a herder-hunter.
7. Although they use eagles, falcons, and hawks, for the sake of convenience, I will speak here only of eagles.
8. *Pood* is an old Russian unit of mass equal to 16.38 kilograms.
9. A herding camp.
10. Maximum weight can reach 220 kilograms (Mongush 2015).
11. A cone-shaped tent covered by birch bark in summer and reindeer hides in winter. *Chadyrlar* is the plural form of *chadyr*.
12. The Chinese colonizers destroyed the tribal social organization in the eighteenth century. The Soviets contributed to further undermining the tribal system (Vainshtein and Mannay-ool 2001). Today, belonging to a particular *söök* is not as important as it was in the past. Only a few elders who still recall the importance of the *söök* may ask about someone's *söök* affiliation.

REFERENCES

Anderson, David G., Jan Peter Laurens Loovers, Sara Asu Schroer, and Robert P. Wishart. 2017. "Architectures of Domestication: On Emplacing Human-Animal Relations in The North." *Journal of the Royal Anthropological Institute* 23, no. 2: 398–416. doi:10.1111/1467-9655.12613_1.

Armstrong Oma, Kristin. 2010. "Between Trust and Domination: Social Contacts Between Humans and Animals." *World Archaeology* 42, no. 2: 175–187. DOI: 10.1080/00438241003672724.

Baskin, Leonid. 2009. *Severnyi Olen'. Upravlenie Populiatsiiami. Olenevodstvo. Okhota.* [Reindeer. Population Management. Reindeer Herding. Hunting]. Moscow: Tovarishchestvo Nauchnykh Izdanii KMK.

Biche-ool, Svetlana. 2007. "Sotsial'naia Struktura Obshchestva i Demograficheskaia Situatsia. [Social Structure and Demographic Situation]." In *Istoria Tuvy* [History of Tuva]. Kyzyl.

Biche-ool, Svetlana, and Aiana Samdan. 2012. "Sovremennoe Sostoianie Olenevodstva Respubliki Tyva" [The Current Status of Reindeer Herding in the Republic of Tyva]. *Novye Issledovaniia Tuvy,* 1: 111–121.

Bogdanova, A. 1962. "Ia v Dal' Gliadel . . ." [I Stared Into the Distance]. *Tuvinskaia Pravda*, no. 59 (March 10).

Bradshaw, John. 2011. *How the New Science of Dog Behavior Can Make You a Better Friend to Your Pet.* New York: Basic Books.

Brandišauskas, Donatas. 2017. *Leaving Footprints in the Taiga. Luck, Spirits and Ambivalence among the Siberian Orochen Reindeer Herders and Hunters.* New York: Berghahn Books.

Brightman, Robert. 1993. *Grateful Prey: Rock Cree Human-Animal Relationships.* Berkeley: University of California Press.

Certified Horsemanship Association. 2017. http://cha-ahse.org/. Accessed: Nov. 13, 2019.

Clutton-Brock, Juliet. 1981. *Domesticated Animals from Early Times.* Austin: University of Texas Press.

Courtiol, Alexandre, Jean-Baptiste Ferdy, Bernard Godelle, Michel Raymond, and Julien Claude. 2010a. "Height and Body Mass Influence on Human Body Outlines: A Quantitative Approach Using an Elliptic Fourier Analysis." *American Journal of Physical Anthropology* 142, no. 1: 22–29.

Courtiol, Alexandre, Michel Raymond, Bernard Godelle, and Jean-Baptiste Ferdy. 2010b. "Mate Choice and Human Stature: Homogamy as a Unified Framework for Understanding Mating Preferences." *Evolution: International Journal of Organic Evolution* 64, no. 8: 2189–2203.

Darzha, Viacheslav. 2013. *Traditsionnye Zaniatiia Tuvintsev. Khoziaistvo, Okhota, Rybalka* [Traditional Occupations of Tyvans. Economy, Hunting, and Fishing]. Vol. 1. Kyzyl: OAO Poligraf.

Davydov, Vladimir, and Veronika Simonova. 2008. "Sobach'e Serdtse: Antropologiia Sobakoedeniia v Postsovetskoi Evenkiiskoi Derevne" [Heart of the Dog: the Anthropology of Dog-Eating in a Post-Soviet Evenki Village]. *Izvestiia Laboratorii Drevnikh Tekhnologii,* no. 6: 213–230.

Dobney K., and G. Larson. 2006. "Genetics and Animal Domestication: New Windows on an Elusive Process." *Journal of Zoology*, 269, no. 2: 261–271.

Fijn, Natasha. 2011. *Living with Herds: Human-Animal Coexistence in Mongolia.* Cambridge: Cambridge University Press.

———. 2015. "The Domestic and the Wild in the Mongolian Horse and the Takhi." In *Taxonomic Tapestries. The Threads of Evolutionary, Behavioral and Conservation Research*, edited by A. M. Behie and M. F. Oxenham, 279–298. Acton: ANU Press.

Gupta, Anil K. 2004. "Origin of Agriculture and Domestication of Plants and Animals Linked to Early Holocene Climate Amelioration." *Current Science* 87, no. 1: 54–59.

Hare, Brian, Michelle Brown, Christina Williamson, and Michael Tomasello. 2002. "The Domestication of Social Cognition in Dogs." *Science* 298, no. 5598: 1634–1636. DOI: 10.1126/science.1072702.

Heptner, Vladimir, and N. P. Naumov. 1998. *Mammals of the Soviet Union, Volume 2 Part 1 a. Sirenia and Carnivora (Sea Cows; Wolves and Bears)*. New Delhi: Amerind Publishing Co. Pvt. Ltd.

Ingold, Tim. 1980. *Hunters, Pastoralists and Ranchers: Reindeer Economies and their Transformations*. Cambridge: Cambridge University Press.

———. 1994. "From Trust to Domination: An Alternative History of Human-Animal Relations." In *Animals and Human Society: Changing Perspectives*, edited by A. Manning and J. Serpell, 1–22. London: Routledge.

Jackson, Linda, and Kelly Ervin. 1992. "Height Stereotypes of Women and Men: The Liabilities of Shortness for both Sexes." *The Journal of Social Psychology* 132, no. 4: 433–445. DOI: 10.1080/00224545.1992.9924723.

Kahn, Paul, and Francis Woodman Cleaves. 1998. *The Secret History of the Mongols: The Origin of Chinghis Khan (expanded Edition): an Adaptation of the Yüan Ch'ao Pi Shih, Based Primarily on the English Translation by Francis Woodman Cleaves*. Boston: Cheng & Tsui Company.

Knight, John. 2012. "The Anonymity of the Hunt: A Critique of Hunting as Sharing." *Current Anthropology* 53, no. 3: 334–355. DOI: 10.1086/665535.

Krasnoiarskstat. 2019. *Pogolov'e Skota i Ptitsy v Munitsipal'nykh Raionakh (Gorodskikh Okrugakh) Respubliki Tyva na 1 Ianvaria 2019 Goda. Ekonomicheskaia Tablitsa, № 1.33.44.1.RT* [Livestock and Poultry in Municipalities (Urban Districts) of the Republic of Tyva on January 1, 2019. Economic Table N 1.33.44.1. RT]. Kyzyl: Krasnoiarskstat.

Kristensen, Benedikte Møller. 2015. "Returning to the Forest: Shamanism, Landscape and History among the Duha of Northern Mongolia." Ph.D. diss., University of Copenhagen.

Laland, Kevin, and Gillian R. Brown. 2002. *Sense and Nonsense. Evolutionary Perspectives on Human Behavior*. New York: Oxford University Press.

Laufer, Berthold. 1917. "The Reindeer and its Domestication." *Memoirs of the American Anthropological Association* 4, no. 2 (1917): 91–147.

Lestel, Dominique, Florence Brunois, and Florence Gaunet. 2006. "Towards an Etho-Ethnology and Ethno-Ethology." *Social Science Information* 45, no. 2: 155–177. DOI: 10.1177/0539018406063633.

Lestel, Dominique. 2006. "Ethology and Ethnology: The Coming Synthesis." *Social Science Information* 45, no. 2: 147–153. DOI: 10.1177/0539018406063632.

Levin, Maksim Grigorievich. 1958. *Etnicheskaia Antropologiia i Problema Etnogeneza Narodov Dal'nego Vostoka* [Ethnic Anthropology and the Problem of the Ethnogenesis of the Peoples of the Far East]. Moscow: Nauka.

Miklósi, Ádam, and Jószef Topál. 2013. "What Does it Take to Become 'Best Friends'? Evolutionary Changes in Canine Social Competence." *Trends in Cognitive Sciences*, 17, no. 6: 287–294. DOI: 10.1016/j.tics.2013.04.005.

Millan, Cesar, and Melissa Jo Peltier. 2014. *Cesar's Way – The Natural, Everyday Guide to Understanding and Correcting Common Dog Problems*. New York: Harmony Books.

Mongush, B. K-O. 2015. "Ekster'er Olenei, Razvodimykh v MUP 'Odugen' Todzhinskogo Raiona." Vypusknaia Kvalifitsionnaia Rabota ["Exterior of Reindeer Bred at the MUP 'Odugen' in the Tozhu district"]. Thesis, Tuvan State University.

Nadasdy, Paul. 2007. "The Gift in the Animal: The Ontology of Hunting and Human-Animal Sociality." *American Ethnologist* 34, no. 1 (2007): 25–43.

Oehler, Alexander Christian. 2016. "Being Between Beings: Soiot Herder-Hunters in a Sacred Landscape." Ph.D. diss., University of Aberdeen.

Pearson, Osbjorn M. 2000. "Activity, Climate, and Postcranial Robusticity: Implications for Modern Human Origins and Scenarios of Adaptive Change." *Current Anthropology* 41, no. 4: 569–607.

Petri, Berhhard. 1927. *Olenevodstvo u Karagas* [Reindeer Herding among the Karagas]. Irkutsk: Tipografiia Izdatel'stva Vlast' Truda.

———. 1928a. *Biudzhet Karagasskogo Khoziaistva* [Budget of the Karagas Economy]. Irkutsk: Tipografiia Izdatel'stva Vlast' Truda, 1928a.

———. 1928b. *Promysly Karagas* [Karagas Economies]. Irkutsk: Tipografiya Izdatel'stva Vlast' Truda.

———. 1930. *Okhota i Olenevosdtvo u Tuturskikh Tungusov v Sviazi Organizatsiei Okhotkhoziaistva* [Hunting and Reindeer Herding among the Tuturskiy Tungus in Connection with the Organization of Hunting Management]. Irkutsk: Tipografiya Izdatel'stva Vlast' Truda.

Salska, Irmina, David A. Frederick, Boguslaw Pawlowski, and Andrew H. Reilly. 2008. "Conditional Mate Preferences: Factors Influencing Preferences for Height." *Personality and Individual Differences* 44, no. 1: 203–215.

Shipman, Pat. 2015. *The Invaders. How Humans and Their Dogs Drove Neanderthals to Extinction*. Cambridge: Harvard University Press.

Simon, Scott. 2015. "Real People, Real Dogs, and Pigs for the Ancestors: The Moral Universe of 'Domestication' in Indigenous Taiwan." *American Anthropologist* 117, no. 4: 693–709. DOI: 10.1111/aman.12350

Stépanoff, Charles, and Jean-Denis Vigne. 2018. "Introduction." In *Hybrid Communities Biosocial Approaches to Domestication and Other Trans-Species Relationships*, 1–20. London: New York: Routledge.

Stépanoff, Charles. 2012. "Human-Animal 'Joint Commitment' in a Reindeer Herding System." *HAU: Journal of Ethnographic Theory* 2, no. 2: 287–312. DOI: 10.14318/hau2.2.015.

Tugolukov, Vladilen Aleksandrovich. 1969. *Sledopyty Verkhom na Oleniakh* [Pathfinders Riding Reindeer]. Moscow: Nauka.

Vainshtein, Sevyan, and Mongush Kh. Mannay-Ool. eds. 2001. *Istoriia Tuvy* [History of Tyva]. Novosibirsk: Nauka.

Vainshtein, Sevyan. 1961. *Tuvintsy-Todzhintsy* [Tuvan-Todzhinians]. Moscow: Izdatel'stvo Vostochnoi Literatury.

———. 1980. *Nomads of South Siberia.* Edited by C. Humphrey. Cambridge: Cambridge University Press.

Vasilevich, Galina M., and Maksim G. Levin. 1951. "Tipy Olenevodstva i ikh Proiskhozhdeniia" [Types of Reindeer Herding and their Origins]. *Sovetskaia Etnografiia,* no. 1: 63–87.

Vasilevich, Galina M., and Maksim G. Levin. 1980. "Proiskhozhdenie Saianskikh Olenevodov: Problema Etnogeneza Tuvintsev-Todzhintsev i Tofalarov" [The Origin of Saian Reindeer Herders: The Problem of the Ethnogenesis of Tyva-Todzhinians and Tofalars]. In *Etnogenez Narodov Severa,* edited by Ilia S. Gurvich, 68–88. Moscow: Nauka.

Weatherford, Jack. 2010. *The Secret History of the Mongol Queens: How Daughters of Genghis Khan Rescued His Empire.* New York: Broadway Paperbacks.

Willerslev, Rane. 2007. *Soul Hunters. Hunting, Animism, and Personhood among the Siberian Yukaghirs.* Berkley: University California Press.

Wolánski, Napoleon, and Ewa Kasprzak. 1976. "Stature as a Measure of Effects of Environmental Change." *Current Anthropology* 17, no. 3: 548–552.

Chapter 2

Predatory Relations of Tozhu Hunter-Herders

Aivaras Jefanovas

This chapter is based on five weeks of ethnographic research with reindeer herders and hunters of the Tozhu District of northeastern Tyva between August and September 2017.[1] In choosing Tyva as a field site, I was driven by the idea to explore the social relationships between Tozhus and animals, especially predators. My interest in this topic had initially been sparked by Russian ethnographer Sevyan Vainshtein's vivid descriptions of the bear honoring ceremony of the mid-twentieth century (Vainshtein 2016, 71–76). This ritual was performed to calm the bear spirit and to deflect possible revenge from the predator, following the kill. During the bear honoring ceremony, the hunters would ritually blame crows for the killing, absolving the Tozhu hunters of their guilt. The bear's skull and paws were then hung on a tree, in the direction of sunset.

Initially I had believed these rituals to be indicators of a true and ongoing human-animal relationship that was based on respect and reciprocity. During my time in Tyva, however, I learned that there were only few overtly performed hunting ceremonies. Nonetheless, certain fragments of the ritual did resurface in the daily routine of hunters who would silently feed the fire, abstain from talking loudly about bears prior to the hunt, or show respect to predators by telling frightening stories and legends.

Interestingly, such subtle ritual forms may have become more popular during Soviet times. The aim would have been to hide rituals from the regime—a regime so powerful that, metaphorically speaking, it was able to suppress not only the shamans but also the spirits (Anderson 2016, 87–91; Pedersen 2011, 52–53). Supposedly, the long domination of the Soviet state, with its ideology of atheism and a planned economy, had weakened the bear honoring rites of Tozhus. However, I would like to think that despite the materialist ideology of past Soviet collective farms, predator honoring activities continue to play a significant role in hunting practices today. These practices are based on personal experience of individual Tozhus, along with recollections of the traditions of their ancestors.

In this chapter I will investigate the sociocultural interaction between Tozhus and predators, based on the daily practices used by reindeer herders and hunters to manage bear and wolf populations. I strive to show how Tozhus and predators, who share a common landscape, interact with each other through adjustment in behavior, by way of cohabitation, and in mutual aggression (See also Brandišauskas 2017, 244–248; 2009, 122–134, 150–154, 252–259). I am particularly interested in how these experiences of each other contribute to Tozhu predator perception.

The favorable ecological particularities of the Tozhu District lend themselves to an exploration of human-predator interaction. Of course, intensive industrialization and mineral resource exploitation during the Soviet era have left deep scars in the taiga, which are visible even today. Abandoned industrial structures and buildings remain, many strewn with rubbish. One can find oil spills, gaping drill holes, and scattered fuel barrels among the remains of rusting geological exploration tools and equipment. In more recent decades, following near total economic stagnation and political upheaval, Chinese and Russian companies have taken up gold and metallurgical mining activities in the taiga. However, their environmental exploitation, as well as their construction of new camps and access roads, has not significantly altered the remote and wild character of the eastern Saian mountains. With their fast-moving rivers, hundreds of lakes, and dense forests, these mountains represent a perfect refuge for predators. Tyvan game managers from Kyzyl, as well as hunters and reindeer herders from the Tozhu District, were certain that there had occurred an increase in predator populations following the collapse of the Soviet collective farms. During these years, wolves and bears had become so numerous because effective control measures were no longer being used. This pattern had continued well into the present, with the current government still being unable to properly manage predator populations.

Tozhu locals also noticed that following the collapse of the Soviet Union, the behavior of predators had changed. It seemed, predators had lost their fear of people, which had encouraged wolves and bears to attack reindeer in their encampments more frequently. Tozhus listed several reasons for these changes in predator behavior. During the Socialist years, helicopters had been flown to shoot wolves from the air, and specialist wolf hunters, or wolfers [Rus. *volchatniki*], had been employed to systematically exterminate wolves. At present few if any of these measures were being implemented. Referring to the present circumstances, reindeer herders argued that in these post-Soviet times of shortage, the government had turned away from its people, and Tozhu locals were now wholly reliant upon themselves.

The socio-economic landscape of the post-Soviet era was marked by political upheaval and the consequences that came with the collapse of state farms, including their subsidies, as well as the loss of local integration into the wider Soviet planned economic system. The resulting socialist economic breakdown of the 1990s had led reindeer herding into a deep crisis. The number of reindeer in Tyva had sharply decreased from 14,000 in 1982 to 1,100 in 2001 (Donahoe 2004, 188–192). However, at the time of my visit, the number of reindeer hovered around 2000 head (cf. Kan 2016, 65). Local reindeer herders recalled that prior to World War II there were said to have been between 20,000 and 30,000 reindeer. The dramatic drop in reindeer following the collapse of collective farms can be attributed to several reasons. Following the collapse of the USSR, state salaries and subsidies had dried up, and even basic supplies and services, including veterinary services and predator control, had disappeared (Donahoe 2004, 188–192). For many herders this post-socialist predicament had resulted in financial ruin, forcing people to slaughter their reindeer for subsistence, or to sell the meat for cash (ibid.). Moreover, reindeer were distributed by the administration of former state farms [Rus. *sovkhoz*], not only to herders, but to all members of the former institutions, most of whom were not reindeer herders and did not know what to do with the reindeer other than to slaughter or sell them (ibid.). Reindeer herders mentioned to me that after the collapse of the state farms, unskilled recipients of reindeer had just left their animals in the taiga without attendance. Consequently, the animals turned feral and many of them fell prey to Tozhu and Russian hunters.

Meanwhile, many Tozhus abandoned their lives as reindeer herders in the taiga and sought accommodation in the villages. However, here too they were left without employment. Most local Tozhus, with the exception of people able to secure government employment, were left with no regular income. This encouraged poaching activities and the illegal trade of musk deer glands, bear paws, and gall bladders on the black market. Many resorted to establishing state pensions on the basis of false declarations of disability, or false war veteran status. To this day, socio-economic instability perpetuates the inability of many Tozhus to participate in and benefit from the larger market economy, which in turn has increased social tension, often leading to open violence (some of which has resulted in murder), suicide, alcoholism, and general strife within communities.

Increased gold and metallurgic mining, combined with the construction of new roads and the establishment of Russian-staffed geological field camps, has led to overexploitation of game animals and fish, exacerbating competition between local Tozhus and Russian newcomers. Furthermore, Tozhu reindeer herders have argued that chemicals emitted from geological camps pollute rivers and lakes, leaving the landscape devoid of game and fish.

Despite these tensions, general lack of revenue for Tozhu locals has encouraged cooperation with the mining camps, where one can trade taiga products, such as game meat and pine nuts, in exchange for cigarettes, tea, and oil. Also, people have been getting free transportation in geologist's vehicles, to and from the village (Donahoe 2004, 257).

In line with the severe socio-economic circumstances of the post-Soviet period, I will discuss here the notion of predation and how it can be elaborated not only in reference to human-animal interactions, but also in the context of relationships among humans, particularly in terms of a combination of aggressive and cooperative actions.

The reindeer herders and hunters I joined in the taiga were local inhabitants of the Tozhu District, who identified as Indigenous Tozhus [Rus. *Tuvintsy-Todzhintsy*] (Donahoe 2004, 64–65). Tozhus are distinguished from other inhabitants by their specific dialect of the Tyvinian language, which contains Samoyed elements (Donahoe 2004, 72–77). Moreover, their traditional economy has been based on reindeer herding and hunting for ages, unlike Tyvans from the central regions of the Republic who have been pastoralists (Sat 1987, 73; Vainshtein 1961, 3). Based in part on these distinctions, Tozhus have been recognized as one of the Russian Federation's "Indigenous Small-numbered Peoples of the North, Siberia and the Far East" since 1993. Around 4,500 Tozhus live in the Tozhu District, the majority of whom reside in the village of Tora-Khem (Mongush 2013, 62–63; Kan 2016, 64).

In pre-revolutionary times, Tozhu nomadic families would roam the Eastern Saian mountains, traversing vast territories in pursuit of reindeer pastures, and following hunting routes, while living in conical lodges [Rus. *chum*], which were temporarily erected in taiga campsites for reindeer herding. After the Revolution, the Soviet regime launched a coercive campaign of sedentarization, which included the construction of villages, the placing of children in boarding schools, and the establishment of collective farms for intensified reindeer husbandry. Since the collapse of these state farms, reindeer herding in Tyva has been organized on the basis of tribal enterprises [Rus. *rodovye obshchiny*], which were intended to fulfill many of the functions formerly performed by state farms. These enterprises organized the productive activities of their herders and hunters, providing equipment and salary (Donahoe 2004, 195–196), while depending on government subsidies that were based on the number of reindeer head within each herd.

During my fieldwork, I visited the nomadic campsite of a group of reindeer herders who were members of the tribal enterprise "Khamsara-Aldyn-Ivi," which owned around eighty reindeer. The enterprise was headed by Svetlana Alekseevna Demkina, a university trained Tozhu leader in her sixties. She had been elected to represent the Association of Indigenous Tozhu-Tyvinians of Siberia [Rus. *Assotsiatsiia Korennykh Narodov Sibirii*

Tuvintsev-Todzhintsev]. Svetlana had devoted her life to developing and maintaining traditional reindeer herding in Tyva. Her parents and grandparents were the descendants of reindeer herders, and she herself had been born at a reindeer camp in the taiga, where she had spent her childhood among reindeer. Svetlana said that the image of reindeer had been imprinted on her mind since childhood. After the Soviet period, and with no money, Svetlana and her husband Vasilii Demkin decided to sell their apartment. With the proceeds they purchased fifty reindeer and established the enterprise. Svetlana told me, that although her husband had been a Russian man, it was he who had inspired her to engage in reindeer herding. Svetlana recalled, the beginning of this activity was very difficult, as she and her husband had jobs in town. Consequently, they could not become full-time herders. Instead, they entrusted the herd to hired hands, which turned out to be irresponsible, losing almost all the animals through negligence. Several of the animals had also secretly been sold for meat. Following this event, Vasilii decided to take herding into his own hands. Having purchased new deer, the couple carefully managed the herd on their own. Unfortunately, Vasilii died in an accident in the early winter of 2008. Traveling from the taiga to the town of Tora-Khem, he attempted to cross a frozen lake with his truck. The ice broke, and Vasilii drowned. After her husband's death, Svetlana was facing a lot of challenges in managing the enterprise and its reindeer. Nonetheless, she had not given up, and one of her three daughters, who also held a university degree, was now helping her with the economic side of the business. The two women argued that some of the main obstacles to reindeer herding in contemporary Tyva were insufficient government funding and a lack of interest in this type of economy. The subsidies that did exist were insufficient to pay herders regular wages, to cover veterinary expenses, or to gain more breeding reindeer in an attempt to increase the herd's genetic diversity.

Svetlana's reindeer were being herded by her relatives, primarily by her brother and her cousin, both named Pavel. Both of the men were experienced elders in their fifties. In summer the two were assisted by their wives, cousins, nieces, and nephews. When I arrived, only two elders were staying at the camp—Svetlana's cousin Pavel and a sixty-year-old elder, Slava, who joined the encampment from time to time.

Pavel told me how, after serving in the army, he had come to work in the *sovkhoz* as a full-time reindeer herder. Today, life in the taiga was different, he said. In his youth there had been many more chums, with up to six families living in each of the herders' encampments. At present, only a few men were scattered across a vast territory. Pavel and Slava were lamenting that today families lived in the village, and the young generation didn't have any motivation to carry out traditional activities in the taiga, as they had nothing to earn here. Even in former Soviet times reindeer herders had stayed mostly in the taiga, and only rarely had they visited the villages to stock up on provisions. Helicopters had transported almost all necessities to the reindeer farms.

Pavel's family lived in a village originally named Ii. It had originally been named after the river by the same name, but at present local Tozhus knew it by its Soviet name: "The First of May" [Rus. *Pervoe Maia*]. This had also been the name of the former state farm. Pavel's wife joined him in the taiga during the warmer months of the year, but his son and daughter rarely visited the camp. Conversely, Pavel would visit his family's house in the village for short periods at a time. In spite of the poor pay and hard work as a herder, this was the life Pavel had chosen for himself. In his words, this way of life had been the best he'd ever lived.

After completing my fieldwork in the autumn of 2017, I received a message from Svetlana, telling me Pavel had traveled for many kilometers to meet up with a group of Tofa herders from Irkutsk Oblast. The Tofas had traveled for eight days to exchange several of their bulls with Pavel. Svetlana had met a Tofa representative at a reindeer herders' conference in Sweden where they had agreed to exchange some of their animals to enhance the genetic diversity in each of their herds. Svetlana had been representing the Association of Indigenous Tozhus at conferences and meetings dealing with reindeer herding issues across Siberia and abroad. She had actively been looking for projects that might provide financial support to reindeer breeders from Siberian regions and foreign countries, for developing and expanding reindeer herding in Tuva. Accounts such as Svetlana's give me confidence that, despite social, economic, and environmental obstacles, reindeer herders of the Saians still have the opportunity to carry on their traditional lifestyle.

PREDATORS AS UNMANAGEABLE TAIGA FIRES

In some regions of the Siberian taiga, predators have become a determining factor in reindeer herding. In some cases bears and wolves may significantly reduce the size of a herd, in other cases they can destroy it entirely (Davydov 2014, 25; Kharinskii 2014, 650–651). During the summer of 2017, while I was visiting with the reindeer herders, bears had invaded their encampment on multiple occasions, taking a total of four calves. Wolves had also attacked the herd in its pasture. They had taken down eleven calves in a single attack. Pavel pointed out that bears were primarily an issue in the summer. Wolves, on the other hand, attacked reindeer all through autumn and winter, as their young were coming of age, requiring more food to sustain themselves. According to Pavel, wolves frequently attacked the herd in late summer, killing large numbers of animals. Unused carcasses would be left behind every time a maternal wolf would teach her young how to hunt. Bears, by contrast, approached the encampment only when compelled by hunger, usually taking no more than a few calves. Summer reindeer pastures were located high in the mountains on open tundra, where wolves were rare visitors, but where bears were a common sight.

Late in autumn and winter, herders followed their reindeer as they descended into river valleys. Here wolves were frequent, and they could become a particular nuisance when the snow formed a crust, allowing the predator to run and easily approach congregating reindeer. However, herders perceived bears and wolves differently, as each of these predators constituted a different threat to Tozhu subsistence. Wolves were generally not seen to pose a direct threat to human lives, but they were feared for their ability to kill large numbers of animals at once. Bears instead took only a small number of calves with each attack, but they could pose a direct threat even to humans. Moreover, Tozhu herders dreaded wolves as killers from a distance, since they were cautious and could rarely be seen. People also perceived them more in terms of roaming packs than as individuals. Bears, on the other hand, were known as individuals. They were holding specific territorial ownership, and people commonly referred to them as masters [Rus. *khoziain*] of the taiga.

When talking about the influx of wolf-packs, Tozhus would often liken them to wildfires. According to the hunter-herders, the wolves were fleeing Mongolia and accumulating in the Saian Mountains. It is interesting to mention that prior to Socialism, there existed the opinion in Tyva that Mongols were deliberately pushing out their predators, driving them into Tyva. This was thought to be an act of revenge for stolen cattle (Darzha 2008, 412). Based on Tozhu associations of wolves with wildfires and Mongol revenge, I suppose that predators were perceived as a natural phenomenon, violent in nature, and provoking a sense of powerlessness in people. Usually this sense of disempowerment was manifest in conjunction with a nostalgia for a better past, that is, the Soviet era, when the state had provided control over predators, whereas at present there was no help from the state. Discourse around predator management commonly surfaced as part of locals' complaints about the current government, blaming it for all sorts of negligence and disorderliness.

Pavel recalled how professional wolfers at every state farm had been charged with trapping and poisoning wolves. These professionals knew how many predators were left within each range, where they migrated, and where they raised their pups. Additionally, they had use of helicopters to help regulate the predator population from the air. In Soviet times, he recalled, fuel had been of no issue, and pilots were able to take off immediately after an incident had been reported. Flying at low altitude, the hunters on board would shoot from the air. According to Pavel, wolves had been managed effectively in those days, drastically reducing their population.

My Tozhu consultants had noticed that since the end of the Soviet era, wolves and bears were no longer afraid of approaching encampments, despite barking dogs and the presence of people. Wolves would prowl around campsites, attacking reindeer. Bears had even broken into cabins, attacking the people inside. In the past, because of the work of state hunters, predators had been so afraid of people that, upon recognizing a rifle on a man's shoulder from a distance, they would have fled immediately.

Where professional wolf hunters were still active, this keen sense of recognition in wolves seemed to persist into the present. According to a specialist wolf hunter, Ivan Oiun of Kyzyl, predators did recognize him in every region of Tyva where he worked. Whenever he set traps, he changed his shoes and altered his scent to camouflage against wolves. He reported: "When I come—that's it—there are no wolves. They leave; there's no predation anywhere. Just after I have left a place, the wolves attack the stock again" (Antuf'ev 2010). However, in eastern Tyva, few hunters had access to the resources required to become a wolf hunter like Oiun.

Local hunters from the village of Tora-Khem noted that in order to succeed in hunting wolves, at least 200 liters of gasoline were required for transportation. Few hunters could afford such experiences. While the authorities had established a bounty for killed wolves, those who had appealed to it, after handing in their furs, often did not receive their money due to bureaucracy and corruption. This left hunters with little stimulus. Pavel also complained that, unlike other pelts, today nobody paid money for wolf skins. Consequently, there was no revenue that might cover the costs incurred by self-initiated predator population management.

People relied solely on themselves, especially in the migratory areas located as far as 200 kilometers outside of Tora-Khem. Here no professional wolf hunters had visited since the collapse of the Soviet Union. According to the herders, hunting wolves required more than access to resources, it took special skills. One needed knowledge about the behavior of predators, about their habits, and one always needed a lot of time. Pavel admitted hunters had to cull entire wolf-packs at once, otherwise their efforts would be pointless, they could even make things worse. If a female wolf survived, it would be a matter of little time before she found another male with whom to mate. The following year's litter would rebuild the pack, and thus they would continue harassing the reindeer. Moreover, surviving wolves were known to take revenge on hunters for killing their fellow pack members. The herders knew that wolf vengeance could lead to the devastation of a significant part of a herd.

The herders primarily used traps and snares to hunt wolves in winter time. These were placed on mountain ridges, where the wind would blow from many sides, taking away human scent. According to Pavel, wolves are extremely sensitive beings, quick to fathom human tricks. It was not sufficient, therefore, to merely place traps on wolf tracks. A hunter had to think ahead of the wolf to overcome it. Before setting a trap, Pavel would remove from it any iron or human odor by boiling it with larch or birch branches. These branches had to be picked from trees in the same area where the traps were to be set.

In hopes of success, hunters sometimes left several traps instead of one. Wolves were known to be capable of gnawing off a single trapped leg, especially if it was frozen, allowing them to get away. A second trap would then catch another leg, securing the catch. Locals would also use sharp iron hooks hidden inside morsels, or sharp edged pieces of measuring tape concealed in meat. Such bites, after being swallowed, would damage the predator's stomach and entrails, resulting in the animal's certain death. However, the Tozhu herders I spoke with thought that traps were not the most effective tools, as wolves were capable of learning and comprehending the matter. Instead they considered poisoning the most effective method, thanks to its lack of odor and the ease with which it spread among animals.

Pavel recalled how, in Soviet times, a barium fluoroacetate mixture was commonly supplied to wolfers by *sovkhozes*. The application of industrial grade chemical poisons to wolves has been outlawed in Russia since 2005. In the past, these chemicals had been extremely effective, as they could exterminate a whole pack of wolves at once. A single wolf had to eat a piece of the bait, spreading the deadly poison with its saliva to the rest of the pack. Reindeer herders report that today only veterinarians had the right to supply alternative poisons, which by comparison to the prohibited barium fluoroacetate had little effect on wolves and could not control their population. However, preparing and setting traps or bait was time consuming, and hunter-herders had to prioritize Herding their reindeer and hunting game for subsistence. Hence, most wolf kills were the result of serendipitous encounters during which hunters simply shot the predator.

Thus, predator hunting was resource and time intensive, and the cost usually outweighed the outcome. Given the lack of stimulus, there were, at the time of my visit, no wolfers in Tozhu District. This situation is the complete opposite to Soviet times, when predator control was highly subsidized by the government. Now reliant upon themselves, locals were applying their own knowledge of predator behavior and hunting skills to inventively outsmart their cunning neighbors. These maneuvers, however, did not amount to what might be considered proper predator population management. Because they were hunted only on occasion, culling attempts merely affected local populations. Perhaps it was for these reasons that reindeer herders sometimes likened predators to taiga fires.

IN THE NEIGHBORHOOD OF PREDATORS: COHABITATION AND RECIPROCAL AGGRESSION

Tyvan reindeer herders experienced imminent closeness to bears and wolves. These predators did not hesitate to approach and kill reindeer, and they also posed a threat to the people at encampments. They usually appeared under cover of darkness, on rainy nights or misty mornings, improving their chances of avoiding armed herders on their incursions into encampments. Anticipating this course of action, people would prepare to fight their predator neighbors, especially at such times. To better illustrate this point, I shall start with my own experience. In early autumn, as the nights were beginning to cool off and snow appeared in the mountains, my time for departure was approaching. On one such evening, as I was washing my road clothes in the cold water of the stream behind our chum, I suddenly noticed Pavel and Slava, rifles over their shoulders, rushing to a nearby lake. On the previous day I had just caught some grayling there. I yelled, "what happened?" They silently pointed to the other side of the lake. A bear was standing there, sniffing the air. Slava shouted to me to release the laika dog, who was loudly barking. After untying the leash, the dog sprinted into the taiga. A while later, the two herders returned to our camp without success, as the bear had escaped to the mountains. Pavel supposed that it would return to the campsite after dark to attack the herd. He concluded: "predators these days don't fear dogs or man."

Later that evening, the herders tethered their dogs closer to the herd, loaded their rifles, and prepared a lamp for the night. I could not shake the feeling that the places I had become familiar with now appeared alien, dangerous, and frightful. It seemed like the taiga had eyes, and it was watching us. The mountains appeared to be concealing some threat behind their cliffs. I could feel the presence of another being nearby. I could read that presence in the stare of the dogs' faces, and from their anxious barking. The herders' nervous movements and whispers only exacerbated this sense of invisible presence. That night, I couldn't sleep at all. I had not been terrified, rather, I felt a respectful kind of fear for the bear. At the same time, I wanted to be sure the herders got their shot, were the bear to reappear. It never did show up that night. I noted the event in my diary as "the bear night." Later, Pavel told me he thought the bear was bored with pine nuts and probably desired the taste of reindeer flesh. Slava had recognized the bear. He said it had visited camp a month earlier, when it had snatched a reindeer calf.

The episode suggests, the herders in the encampment and the predator in the hills above, perfectly knew about the presence and intentions of each other. It would seem that the bear had kept its distance from the campsite, knowing that the barking dogs at the camp would signal its approach to the herders. Interestingly, while the dogs were barking almost incessantly throughout the night, Pavel and Slava—unlike myself—awoke only when the barking became particularly intense. In other words, when the bear made an attempt to approach closer, the herders were immediately aware of it. But the bear had stayed near the mountains where it likely waited for darkness, anticipating the rainy night to silence its movement. At least such a strategy might have enhanced its chance at obtaining a reindeer and escaping to safety. It seemed, the bear had stayed above the camp for most of the night. Yet it never did enter, likely because it had been a moonlit night.

The herders responded with ease to the actions of the bear. It seemed, they were accustomed to the habits of their neighbor. Moreover, the herders knew the bear personally and perceived him more like a persona non grata than a damned enemy pest.

While the reindeer herders were aware of their inability to fight all predators, predators seemed equally conscious of the risks of being shot. This mutual awareness led to a kind of human-predator cohabitation, which was sustained by reciprocal respect of territorial ownership, and it involved non-verbal communication. Examples of such communication were the warning shots the herders fired into the air, as well as territorial markings left on trees. Herders would peel the bark from larch trunks, arguing, bears could recognize the marks of people and that they would generally avoid such places. The herders believed that bears could easily understand this communication, as the places belonging to bears were similarly marked. Bears would scratch or knock down trees, leaving bite and claw marks on the trunks to mark their own territory. The bears knew to position their markings higher up to signal their strength and stature to other animals and to people who might compete with them over territory (Brandišauskas 2017, 156; Simonova 2011, 229).

In spite of this two-way communication, Pavel admitted that his efforts to mark several trees were often ignored by bears. Instead, bears would fearlessly enter the campsite. He had even seen bears leaving scratch marks on the posts of human-built forest storage platforms [Rus. *labaz*], as if to assert, according to Pavel, their mastery rights and aggressive intentions. Other times, hunters would hide on a platform, waiting for an opportune moment to shoot an approaching predator that had been attracted to the spot with a lure, such as the head of a moose. These encounters seem to suggest that predators sought to explore the vigilance of people, while attempting to cross the boundaries of human domestic places. This is further enabled in that nomadic campsites are not separate or fenced off places. They are open constructions, built for temporal living and herding. Hence, the domestic (conventionally seen as a cultural sphere) and the wild (often associated with notions of untouched nature) face each other, constituting a shared landscape where humans and predators interact (Anderson 2017, 142).

However, by crossing domestic boundaries—an act that often resulted in the predation of livestock—predators engaged in a kind of reciprocal aggression with people, who were prepared to pursue and destroy the offenders. In one case, a bear had become so "cheeky," according to the herders, that he had joined the very migratory movements of the herd. The bear had been seen many times walking behind or on the side of the herd, preying on calves or weak reindeer. In this case the herders aimed to shoot the bear right away, as the purpose of its perpetual proximity was evident. However, such decisions were not made lightly. Stalking and killing predators was risky, as Pavel explained. Bears were also known to be angered, chasing down people in retribution for having been wounded, or for having crossed their tracks. According to Pavel, a bear would eventually find the hunter and attack him for having been challenged in its territory.

Russian zoologist, Mikhail Krechmar (2005), describes bears as predators who, being injured or disturbed, may turn to attack their foes. The bigger the animal, the stronger its sense of entitlement to a given territory may be. Large males are said to be so self-confident as the masters of a given place, that they will turn even on an armed person in an effort to drive out the intruder (Krechmar 2005, 235–258). Tozhu reindeer herders believe that bears, as a highly sensitive animal, are able to detect a person's scent, including the odor left on a trail. A bear will, for instance, recognize the scent of a hunter who has touched its pawprint to determine the freshness of the trace. Moving within its area, a bear constantly checks on its own territorial markings (Brandišauskas, 2017, 156–157). Supposedly, when a bear notices a hunter's tracks, it becomes cautious and prepares for an encounter, knowing that humans will follow it with the intention to kill. Intuitively, the bear is aware of this.

Consequently, hunters were keen to hide their intentions to hunt bears and avoided any talk about them. A hunter would not even dare think badly about bears. It was understood that a bear could hear the words, even if whispered from many kilometers away. Similarly, a bear could catch a person's thoughts (Brandišauskas 2017, 153–159). I will provide here an example from fieldwork I have been conducting further north in eastern Iakutia, where I asked a Iakutian hunter about bear funeral traditions. The man interrupted me, saying, "It is not customary to talk about the bear when outside, because he can catch the vibrations coming through the ground. You can whisper about it later in the house, as the floor isolates the words."

However, bear hunting entailed not only risk of imminent physical attack. It was the bear's soul, which could seek revenge, potentially causing people all kinds of harm. According to Soviet ethnographer Sevyan Vainshtein, Tozhu hunters ritually honored the bears they killed so as to calm their spirit. Vainshtein had observed such bear ceremonies among Tozhus as late as the mid-twentieth century (Vainshtein 2016, 71–76). During the ceremony, a group of hunters would boil the head of the bear they had hunted. The eldest among them would then take the first piece of meat from it, passing the head on to the others, pronouncing: "taste the Toray head." He would then imitate the sound of a crow—"chuuk." The others would respond by saying: "It's not we who killed you, Toray, the crows (or Russians) did it." When the meat was eaten, the eldest hunter would take the bear's skull and pronounce, "Let's color the Toray's head with charcoal." The hunters would then mount the blackened bear skull, as well as the bear's paws, on a cedar or young birch tree, with the skull pointing westward (Vainshtein 2016, 72–73).

One of the first things I noticed upon arriving at the reindeer camp were two bear skins of bears, which had been tossed behind the chum. I asked Pavel to tell me how these bears had been shot. He pointed out that a month earlier bears had come near the chum, and the men had shot them on the spot. Seeing only the skins, I was wondering where the herders had put the skulls, paws, and bones. Pavel told me, a man had come from the village, and he had taken these parts to town. When I asked Pavel whether they had performed the honoring ritual, he just kept silent. At first, I felt disappointed because I had hoped to witness a bear ceremonial of the type described in such detail by Sevyan Vainshtein.

Upon my return to the city of Kyzyl, I met a Tozhu elder who had been a professional bear hunter earlier in his life. People could always tell his profession by the many bear skulls that were mounted on the fence poles around his house in the village of Tora-Khem. He was proud to have held such a job, mentioning that only experienced hunters dared to kill bears. Such people had always kept the tradition of honoring the bear. I asked him to tell me about the bear ritual. He told me that many locals no longer observed this ceremony. However, it really depended on the individual. Even those who skipped the ritual would often still express their thanks to the spirit of the taiga for the gifts they had received.

Encouraged by my interest in Tozhu hunting traditions, the elder told me about some of the principles he followed to show respect to the spirits that dwell in taiga places, and also to bears. Here are several examples: "When I am hunting with my companions in taiga, we sleep under an open sky, lying side by side around the fire, but never in a straight line against it. We believe that disrespect of this principle angers the *khoziain* [Rus. for master], and he may drop a trunk across the sleeping people." "Moreover, I never build huts in hunting grounds, as the Russians do, because I can see how the garbage accumulates around such huts, and in my mind such disrespect of taiga places irritates the khoziain." "When going on a hunt, especially on bears, I never take companions who smoke or talk too much, not even those who snore during the night. I believe any strange noise or disturbance—even the smell of a cigarette—can lead to misfortune, because the bear can sense from a distance that a person is planning to kill him, and he will turn to attack the person in exchange."

Not all people were keen on hunting bears. Even in Tora-Khem I was told, most people were afraid of coming across bear tracks in the taiga. An example of the trepidation called for by bears comes from a middle-aged Tozhu hunter from the village:

> While I was alone [in the taiga], I made a shot and wounded [a bear]. As I had feared, the bear took off into the taiga, and I had to follow his bloody tracks. It took a lot of time to trace him and finish him off—I was horrified. Afterwards, my body continued to tremble and I cried, alone. I couldn't even show myself to people. I couldn't talk at all. I stayed in the taiga for a while.

Other fearful stories of bear encounters were circulating, some of which seemed to be told in part for their entertainment value, and in part for their educational worth. Here is an example from my field diary:

> On a rainy cold late night, with a humid wind blowing into the chum from the holes beneath, Pavel turned his back to the fading warmth of the hearth, and whispered that autumn was coming. Slava cursed the Chinese-made stove for its lack of warmth, throwing on more firewood. Then he rolled some tobacco into a piece of old newspaper and silently smoked in the corner. Suddenly the dogs started barking, and the two men jumped outside with their rifles on the ready. When they came back a minute later, I asked if it had been a predator. Pavel smiled and said, "no, it was just . . . ." A while later he tried to tease me with this story: "Once a bear invaded a chum. Several men were sleeping inside. They grabbed their knives and stabbed the beast." Then Pavel and Slava pointed to my knife and smiled again. "So," they said, "a bear can enter the chum at night, but your knife is enough to stop it." Pavel explained that when a bear is about to attack, it will often stand on its two hind legs. "If you throw something at the bear, it will try to catch it, standing erect with its paws stretched out. This is when you stab it right to the heart."

Pavel's advice was followed by a legend to further illustrate the power balance between bears and people:

> Once a hunter, armed only with a knife, crawled into a bear den. In the dark, he groped for the pulsing heart of the animal. When he had found it, he stabbed it to death. But he had not noticed that another bear was hibernating nearby. Awoken by the barking of the dogs outside, it killed the man. Later, other hunters accidentally found the head of the missing man inside an eagle's nest. The eagle had picked up the man's head from among the remnants of the bear's table, and brought it in for its chicks to eat.

While this legendary tale speaks of the power of bears, as much as it points to the relative vulnerability of people, other accounts warned of the high level of intelligence found in bears. I was told, for instance, that when it gets colder, a bear may enter a river or lake to submerge its fur in water. The bear's wet coat will freeze quickly, forming an ice shield, which repels smaller bullets, including buckshot. Although the veracity of these and other accounts is difficult to establish, they indicate the high regard and deep respect hunters had for bears in the eastern Saians. It is not difficult to appreciate why Tozhu hunters used to engage in ceremonies aimed at honoring the bear spirit and fending off misfortune. Even if locals no longer observed many of the bear-related traditions, such as properly mounting the skull of a hunted bear, human-bear relations were still grounded in deeply respectful perceptions of bear personhood. Oral accounts, and the retelling of frightful legends, continued to guide the life and subsistence of Tozhus in a landscape intimately shared with predators.

Sharing a common living space, Tozhus and their predator neighbors seemed well aware of each other's habits and intentions. Where their divergent intentions collided, bear-human interaction became entwined in a pattern of reciprocal aggression. In most cases the resulting tension between mutual reverence and conflict did not seem to cause contradiction in the minds of Tozhu hunters, who always had a practical explanation for bear-human encounters. "We fear bears and we respect their territorial ownership, but if they hang around posing threat, we are keen to destroy them as opportunity lends itself. This is, of course, a dangerous task because a challenged bear may decide to hunt down a person in exchange."

PERCEIVING THE PHENOMENON OF PREDATION: IT WILL EVENTUALLY FIND YOU AND TAKE YOU

People can become the victims of carnivorous spirits, which, akin to predators, may pursue and harm them. Tozhus refer to such spirits as *aza*. In the folklore of South Siberian Turkic tribes, to whom Tozhus are related, *aza* are perceived as the evil spirits or demons of animal-like or anthropomorphic appearance (Oinotkinova 2018, 40). *Aza* spirits seek to kidnap a person's soul, bringing illness or misleading the person. According to Pavel and Slava, "an aza will hunt you down until it finds you and takes you: it doesn't matter how long it takes—a day or a year—it will eventually find you."

Aza spirits could also inhabit certain places. While walking, together with Pavel and Slava, to locate reindeer scattered in the taiga, I pointed to a particular mountain top and told the men I was going to climb it. The two stopped me for a moment, warning me not to go there because their elders had believed that by going there, resident evil spirits could be disturbed, bringing snow, blizzards, rain, lightning, or general misfortune. To illustrate their point, the herders told me the story of a group of Russian geologists who had come from Moscow. Their expedition had led them to a 'bad' place on the mountain—a place where Tozhu elders never walked out of fear of evil spirits. According to Pavel, every one of the Russian visitors eventually died. Not one returned to Moscow alive.

I was also told that *aza* not only dwell in the landscape, but they could also inhabit animals. A man may hunt such an animal without recognizing the harmful spirit inside it. Disturbed by the hunter, the *aza* may be angered, turning against its pursuer. Sooner or later the spirit will jump into the hunter, after which he will fall ill or even die. However, sometimes a hunter is able to discern the presence of an *aza* in an animal. He can then safely avoid it. A bear indwelled by an *aza* is of an unusually bright coat color, or nearly black. Such animals should always be avoided.

I suppose that in local understanding the phenomenon of *aza* expresses the idea of predation. *Aza* are perceived as indwelling certain places, such as mountain ridges, peaks, or even animals. It is worth mentioning that in Tyva there also exists the notion of spirits as owners of places [Rus. *khoziaeva mestnosti*]. These spirits are known as *cher eezi*, and they too are the owners of wild animals. However, it seems that *cher eezi* are perceived by Tozhus differently from *aza*. Hunters are keen to establish reciprocal relationships with *cher eezi*, because an honored and respected spirit may give an animal in exchange. Before venturing on a hunt, the hunter always feeds the *cher eezi* by tossing a morsel of food into the fire, even if all he has is dried bread (Donahoe 2004, 118). Yet, there were certain places in the taiga where people would not hunt, cut down trees, light fires, or camp overnight, as this would disturb the *cher eezi*, bringing trouble or even death to the people. It was also believed that Tozhu hunters must observe proscriptions against killing white animals, such as white elk, white moose, and white bears. Any completely white thing was said to be the *cher eezi* (Donahoe 2004, 120–122). Thus, *cher eezi* were perceived as ambiguous beings. They were partners of mutual exchange, but they could also be revengeful punishers of disrespectful conduct. *Aza* by contrast were always associated with provocation, aggression, and revenge, which interestingly was similar to the herders' concept of predation. If a bear, also known as master of the taiga, had been challenged by a human crossing its path, it would pursue the person until it had overtaken them. In other words, if one fell into a prey-predator relation, one would be overtaken. This seemed to apply to both tangible and intangible predatory encounters.

French anthropologist, Charles Stépanoff (2009), writes about North Siberian Indigenous beliefs regarding spirits that consume human flesh. Koriak kala spirits devour the human liver, and so do Chukchi kêly spirits. Likewise, malevolent spirits will enter the bodies of Yukaghir men and women, eating their internal organs and causing all kinds of disease (Stepanof 2009, 285). The spirits' appetite for human flesh is described as a hunt in which the spirits are the hunters, and humans are game. In other words, from the spirits' perspective, humans are prey. The fact that a spirit may see a person as prey, and not as a flower or a cloud, is constitutive of the prey-predator relation (Stépanoff, 2009, 293).

The concept of predation, as I have described it here, may also be an expression of the human relation to a post-Soviet environment, marked by chronic shortage. Here people feel incapable of managing their own well-being. The predator role is embodied by a state of abandonment, a sense of fear, of disorder, and shortage of state subsidy. A similar sense of post-Soviet predation is found among Evenki hunter-herders of Zabaikal'ia, who believe that many old, abandoned, and tragic places in the taiga have become the habitation of vengeful, malevolent, and cannibalistic spirits, which symbolize Evenki experiences of unpredictable and dangerous contemporary socio-economic landscapes (Brandišauskas 2017, 101–102; 2018).

In like manner, the nature of many human interactions in post-Soviet environments could be described as predatory. As Brandišauskas (2017, 189–217) has shown, such interactions can be based on a combination of aggression and cooperation, as I will intimate below. I would like to parallel this notion of predation with human-predator interactions, which are positioned in shared landscapes, but which are determined not only by acts of reciprocal aggression, but also by cohabitation. Combining the two, Tozhu locals are able to increase their own opportunities for getting ahead. Hence, this combination of asserting aggression and of soliciting cooperation may help us better understand how the concept is locally perceived and lived out.

During my fieldwork in Tyva, Tozhus and Russians told me stories of predatory behavior in people. Particularly herders would argue that nowadays everyone had turned into greedy commercial traders [Rus. *kommersanty*], who produced nothing, but instead bought products cheaply from government facilities, and then resold them at very high prices. In the past, such business opportunists [Rus. *spekulianty*] had been condemned, but today they were heroes.

To further illustrate what I mean by predatory-cooperative relations, I will summarize an interesting incident that was related to me by a Russian worker at a remote metal ore exploration base deep in the taiga. Located 150 kilometers from Tora-Khem, the base was accessed via a newly built gravel road. Three men had been left to watch over the base one day, when five Tozhu locals arrived in an old truck, demanding gasoline. Taken aback by the brazen demand of people, who seemed to have come to the taiga without gasoline only to demand it of others, the Russian workers angrily rejected their visitors' request. Having been turned down, the Tozhu visitors threatened to shoot. Following a dispute, the visitors challenged the workers to sit down with them and drink some moonshine. The Russians agreed.

According to the Russian worker who told me the story, the visitors had kept refilling everyone's glasses in an attempt to provoke a conflict. They had tried to single out the weakest individual among the workers, the one who would first resort to physical aggression under the given pressure. In doing so, the visitors had acted like a pack of wolves who identify the weakest member in a herd and then attack it. If the resulting altercation were to end in death for someone, it would not have been the visitors who had instigated the attack. Paradoxically, this way of blaming others for the death of one's own victim corresponds to the principle of the bear ceremony in which Tozhu hunters blamed crows (or Russians) for the death, thus calming the bear spirit and deflecting its revenge.

I suppose, the Tozhu hunters and their Russian hosts knew perfectly well of each other's intentions, and that each party had adjusted their actions accordingly. The group of visitors had known about the three isolated workers left at the base. Knowing the workers were outnumbered, the hunters had felt confident to impose their aggressive plan to obtain free gasoline. Knowing also that an overt attack would have immediate legal repercussions, the Tozhu visitors attempted a provocation. Conversely, the Russian workers were aware of these circumstances and prevented the scheme from taking effect.

Paradoxically, the same workers also engaged in trade with local Tozhu visitors on a regular basis. Such cooperative interactions included exchanging game meat, reindeer milk, fish, berries, and pine nuts, all of which were traded for cigarettes, gasoline, or other goods from the geologists' base. Both parties also exchanged information about game movement, hunting knowledge, and fishing skills. They would also exchange with each other news from village and city. Some Russians, however, argued that despite this seeming cooperation, Tozhu locals were greedy and unwilling to share anything for free. Some were said to have taken a stove and gloves from a vehicle that had belonged to the base.

The idea that consumption requires reciprocity is widely spread in Siberia. In other words, cooperative interaction is understood to engage in mutual exchanges (Brandišauskas 2017, 113–115, 245–250; Davydov 2014a, 2014b; Wilersev 2007, 35–42). For local Tozhus, taking things from the base could be understood as congruent with the notion of reciprocal engagement, in which all things may be perceived as a shared resource. Historically, a Tozhu hunter was obliged to surrender whole or part of his prey to any stranger met in the taiga, who would pronounce the word "*uzha*." Sevyan I. Vainshtein observed this tradition as late as the mid-twentieth century (2016, 69–70). From the perspective of the Russian workers at the geologists' base, the incident may have seemed more like an act of non-equal exchange or theft. It could also be argued, given the Tozhu predicament of scarcity and economic instability in this post-Soviet landscape, that some were manipulating old principles, simply to get what they needed.

It seemed to me that most local decisions were a matter of constant adjustment; people seemed intuitively aware of their circumstances, fluidly taking on roles as aggressors, and as cooperators (e.g. as exchange partners). These ideas resonate with Morten Axel Pedersen's investigation in post-socialist Mongolia. Pedersen demonstrates that in rural settlements, the official state lacked obvious power. Instead, the strongmen, or atamans, violently dominated in local communities. However, atamans' tactics were by no means always unwelcome. They demonstrated a peculiar switching mechanism that allowed them to turn from their bandit source of authority to domination by way of subtle words. The author compares the ataman with a shamanic persona for his capacity to enter into states of being others cannot follow, and to be in control of what he is doing while he is in such a state (Pedersen 2011, 68–71).

However, not always could the actions of Tozhu locals be interpreted as intuitive manipulation between states of being. Aggression could also erupt spontaneously when stimulated by alcohol. To illustrate a sudden shift from cooperation to aggression, I will recount my own predicament as an outsider, falling into what may be considered prey-predator relations with a group of locals. In my search of the reindeer camp, a Tozhu hunter from Tora-Khem had kindly agreed to drive me deep into the taiga with his truck. As we were crossing Chamsara River, the truck got stuck in the middle of the river, and it almost sank. After a while, two motor boats arrived, both loaded down with local hunters. It was suggested I pay money for one of the boats to return to the village where a second truck could be hired to salvage the first. Most of the passengers on the boats were drunk, leaving me uncertain whether they would simply vanish with the money.

My driver assured me he knew the men, and they were obliged to help us, or else they would lose face in the village [Rus. *oni ne smogut posmotret' muzhikam v glaza*]. After agreeing to pay, one boat returned to the village, while the other took us to shore. We settled with the middle-aged group of hunters, all of whom looked quite friendly, sharing their food with us. After midnight the lights of another truck appeared on the other side of the river, and with it had arrived the second boat and its crew of younger hunters. After the sinking truck had been salvaged, I found myself in the company of roughly twenty Tozhus sitting round the fire. More alcohol was consumed, and the situation was changing to one in which I was provocatively being interrogated about my nationality and about what I was doing here. My answers only seemed to further incite some, and before long, I was being pressed for alcohol and for more money. The men would not let my driver and I go, and feeling responsible for my safety, he handed them the money I had paid him for his services. Eventually the men let us go, and my driver assured me they would not have harmed us for fear of revenge from his brothers.

The river incident serves as an example of the ease with which cooperation can turn to aggression as a matter of perspectival shift; a shift that is aided by alcohol. In other words, the alcohol helped mediate the transition to a predator-prey relational status in which one agent came to see the other as prey. This is not an unusual phenomenon, and I would like to draw a parallel here to Stépanoff's description of Siberian cannibal shamans. Having temporarily acquired the perspective of a human devouring spirit, the shaman comes to see people as prey (Stépanoff 2009, 286–288). He will now bite—or even consume—the people around him, devouring their souls, or even their raw blood and organs. This is to say, his perspective is informative of his relational status in a predator-prey constellation.

Interestingly, Stépanoff points out that the notions of prey and predator are relational by definition, and that one cannot exist without the other, or, rather a person is simultaneously part of a multiple whole. Certainly, this idea resembles what Marylin Strathern has referred to as merographic, a viewpoint that anything may be a part of something else (1992, 72–73). As much as a man can be both father of a son and son of a father at the same time, so the shaman's multiple relations do not contradict each other (Stépanoff 2009, 289–290). From a Tozhu point of view, it may be argued, there is no inherent contradiction between cooperative and predatory dispositions, especially where alcohol is allowed to ease collective perspectival shift. Much like the shamans described by Stépanoff, Tozhu locals, aided by drink, would lapse into an aggressive predatory frame of interaction, while later regaining their normal state of mind. This is also what the shaman experiences: putting on his costume, he assumes the perspective of a predatory spirit. When taking it off again, he returns to his untransformed self (Stépanoff 2009, 287–288).

Of course, it would not be justified to simply accuse Tozhus of being brutal alcoholics, or framing the incident as a unique characteristic of their society. However, much like Pedersen talks about post-socialist Mongolia, even a generally stable community that is subject to chronic unemployment and social instability may at times suffer from chaotic experiences with alcoholism, violence, and aggression (Pedersen 2011, 53–60).

I would argue that, perhaps, alcohol eases perspectival shift to aggressive actions, which is a quality that has been integral to Indigenous subsistence in the taiga, independently of drunkenness. In a sense, Tozhus' interaction with the taiga has always been based on taking advantage of opportunities in sometimes harsh and adverse environments. It is not surprising that this often meant acting aggressively against others, whether fighting with bears or Tsarist Cossacks, resisting Bolsheviks or struggling in a savage market economy.

As we have seen throughout this chapter, the notion of predatory relations can extend far beyond human-animal interactions (See also Brandišauskas 2009, 233–234, 247; 2017, 224–226, 240). In fact, it can be used to examine the ways in which people coped under the challenging socioeconomic circumstances of a post-Soviet state. This struggle may take on cooperative or aggressive forms, without relying on overt boundaries between the two. It may be said, my Tozhu consultants fluidly moved between these relational frames, reflecting their intuitive awareness of changing circumstances.

However, interaction based on a combination of aggressive and cooperative relationships between Tozhus and other nationalities, particularly in the case of the Russians workers at the geological exploration base, demonstrates clear affinity with the human-predator relations described in the first half of this chapter. Here too, aggressive action was part and parcel of long-term cohabitation in a shared landscape. In other words, many Tozhus' relationships with their human neighbors maintained a striking resemblance to the intuitive methods around which predator-prey social interactions were patterned in more-than-human contexts of the deep taiga.

NOTE

1. My field research for this article has been funded by the Faculty of Philosophy of Vilnius University and the Lithuanian Research Council (EU structural funds) as part of my doctoral studies. This research would not have been possible without my PhD supervisor, prof. Donatas Brandišauskas. I thank Svetlana Alekseevna Demkina, representative of the Assoc. of Indigenous Tozhu-Tyvinians of Siberia and the Tozhu herders and hunters for their collaboration and support.

REFERENCES

Anderson, David G. 2016. "Shamanistic Revival in a Post-Socialist Landscape: Luck and Ritual among Zabaikal'e Orochen-Evenkis." In *Landscape and Culture in Northern Eurasia* edited by Peter Jordan, 71–96. New York: Routledge.

———. 2017. "Humans and Animals in Northern Regions." *Annual Review of Anthropology* 46 (1): 133–149. doi:10.1146/annurev-anthro-102116-041556.

Antuf'ev, Evgenii. 2010. "Volchii Bilet [Wolf's Pass]." *Tsentr Azii*, no. 15. http://www.centerasia.ru/issue/2010/15/3359-volchiy-bilet.html. Acessed: Nov. 11, 2019.

Brandišauskas, Donatas. 2009. Leaving Footprints in the Taiga: Enacted and EmplacedPower and Luck among Orochen-Evenki of Zabaikal Region in EastSiberia. PhD diss., University of Aberdeen.

———. 2017. New York: Berghahn Books.

———. 2018. "Gluttons of East Siberia: Spirits, Poachers and Cannibals in Evenki Perceptions." In *Human-Environment Relations: Skills, Knowledge, Emotions and Politics in Siberia and China (Evenk and Even Peoples).*

Darzha, Viacheslav. 2008. *Traditsionnye Muzhskie Zaniatiia Tuvintsev. T. I.: Khoziaistvo. Okhota. Rybalka.* [Traditional Men's Occupations Of Tuvans. Vol. 1: Household. Hunting. Fishing]. Kyzyl: Tuvinskoe Knizhnoe Izdatel'stvo.

Davydov, Vladimir. 2014a. "Fishery in 'Free Spaces': Non-Compliance with Fishery Regulations in a Northern Baikal Evenki Village." *Polar Record* 50 (4): 379–390. doi:10.1017/s0032247414000163.

———. 2014b. "Coming Back to the Same Places: The Ethnography of Human-Reindeer Relations in the Northern Baikal Region." *Journal of Ethnology and Folkloristics* 8 (2): 7–32.

Donahoe, Brian. 2004. "A Line in the Sayans: History and Divergent Perceptions of Property among the Tozhu and Tofa of South Siberia." PhD diss., Indiana University.

Kan, Valeriia. 2016. "Etnosotsial'nyi Profil' Tuvintsev [Ethnosocial Profile of Tuvans]." *Novye Issledovaniia Tuvy*, no. 2. https://nit.tuva.asia/nit/art%20icle/view/94/183. Accessed: Nov. 11, 2019.

Kharinskii, Artur. 2014. "Sovremennoe Olenevodstvo Severobaikal'skikh Evenkov." In *Sibirskii Sbornik-4: Grani Sotsial'nogo: Antropologicheskie Perspektivy Issledovaniia Sotsial'nykh Otnoshenii i Kul'tury* edited by Vladimir Davydov and Dmitry Arzyutov, 648–663. St. Petersburg: Muzei Antropologii i Etnografii im. Petra Velikogo (Kunstkamera) RAN.

Krechmar, Mikhail. 2005. *Mokhnatyi Bog* [Hairy God]. Moscow: Izdatel'stvo ID Bukhgalteriia i Banki.

Mongush, Marina. 2013. "O Dvukh Ekspeditsiiakh v Todzhu [About Two Expeditions to Todzha]." *Novye Issledovaniia Tuvy*, no. 1. https://www.tuva.asia/journal/issue_17/5968-mongush.html. Accessed: Nov. 11, 2019.

Oinotkinova, Nadezhda. 2018. Demonim "Aza" v Iazyke i Fol'klore Tiurkov Iuzhnoi Sibiri [Demonym "Aza" in the Language and Folklore of Turkic Peoples of Southern Siberia]." *Sibirskii Filologicheskii Zhurnal*, no. 2: 40–52.

Pedersen, Morten Axel. 2011. *Not Quite Shamans: Spirit Worlds and Political Lives in Northern Mongolia*. Ithaca, NY: Cornell University Press.

Sat, Shuluu. 1987. *Tyva Dialektologiia* [Tyva Dialectology]. Kyzyl: Tyvanyng Nom Ündürer Cheri.

Simonova, Veronika. 2011. "Living Taiga Memories: How Landscape Creates Remembering among Evenkis in the North Baikal, Siberia." PhD Diss., University of Aberdeen.

Stépanoff, Charles. 2009. "Devouring Perspectives: On Cannibal Shamans in Siberia." *Inner Asia* 11 (2): 283–307. doi:10.1163/000000009793066451.

Strathern, Marilyn. 1992. *After Nature: English Kinship in the Late Twentieth Century*. Cambridge: University Press.

Vainshtein, Sevyan. 1961. *Tuvintsy-Todzhintsy. Istoriko-Etnograficheskie Ocherki* [Tuvans – Todzhns: Historical and Ethnographic Essays]. Moscow: Izdatel'stvo Vostochnoi Literatury.

———. 2016. *Zagadochnaia Tuva* [Mysterious Tuva]. Abakan: OOO "IPP 'Zhurnalist.'"

Willerslev, Rane. 2007. *Soul Hunters: Hunting, Animism, and Personhood Among the Siberian Yukaghirs*. Berkeley: University of California Press.

Chapter 3

Emplacing Herder-Horse Bonds in Ak-Erik, South Tyva

Victoria Soyan Peemot

MULTISPECIES INTERRELATIONS IN TAME LANDSCAPES

It is common in Tyva to ask a new acquaintance where he or she is from. I answer with one word: "akaa."[1] It says that I am a Soian from Ak-Erik, Tes-Khem, a southern province of Tyva republic on the border with Mongolia. The Ak-Erik people address their elder men as *akaa* [Tyv. for elder brother]; "the rest of Tyva" uses the term *aky* [Tyv. for elder brother].[2] As a result of a metonymical transfer, "akaa"[3] has become a vernacular term of (self) identification for the Soians who live south of the River Tes.[4] *Akaalar*[5] pride themselves as gifted equestrians with good horses that graze on steppes of especially nutritious *hot*-grass,[6] and where open space is particularly suitable for horse racing. Born in Ak-Erik and brought up by my paternal grandparents, I used to share the Ak-Erik herders' life-world; now I try to be a part of it in a different way—through collaborative research on mutual belonging with landscapes and nonhumans.

This chapter examines how herder-horse bonds are emplaced in landscapes. I argue that bonds are emplaced[7] through interrelations between humans, nonhumans, and landscapes. In this study I emphasize tame landscapes of human-horse interactions—the yurt's interior, the camping site's territory [Tyv. *kodan*], and the seasonal camping site [Tyv. *khonash*]. Horse-yurt symbolism implies the horse's high symbolic value and a gendered approach to horse husbandry among Ak-Erik herders. The duration of time that various domesticated animals spend in the kodan, and the specifics of their enclosures in the kodan territory, suggest horses have more autonomy compared to other livestock. The seasonal khonash campsites tell stories of different generations of multispecies communities and the cycle of their lives.

First, I will introduce the Tyvan concept of the multispecies unit, known as *aal* [Tyv.]. The concept of *aal* resides at the core of every herders' perception of landscapes as shared with nonhumans. Then, I will explain the status of the horse as livestock of the land in order to demonstrate how horses relate to landscapes differently from other domesticated animals. Finally, I will show how herder-horse bonds are emplaced in the yurt, the immediate kodan territory, and beyond the kodan, in the landscapes of seasonal khonash camp sites.

ON THE CONCEPT OF EMPLACEMENT

Ak-Erik herders use narrowly specified emplacement concepts—*aal, kodan,* and *khonash*—that carry knowledge on the specifics of human-nonhuman bonds. The ability to read inherent meanings in landscapes keeps multispecies relationships in a balance. I use the term "landscape" in this study because Indigenous concepts do not translate into English coherently. A literal translation of "landscape" in the Tyvan language is *boidus churumaly*, "a picture of nature." The term suggests an appreciative and observational approach to the land. However, the herders do not talk about static "pictures of nature"; instead, they talk about their interrelated birthplace, homeland, seasonal campsites and pastures, the territories of actual campsites, and abandoned campsites. The constituents of landscapes have names and stories of their own. Some of them are inhabited by their own masters—[Tyv. *eezi*]—nonhuman beings who have their own agency and engage with animals and humans. In this study, "landscape" acknowledges human-nonhuman interrelations. The notion of the "engaged landscape" is inspired by anthropologist Caroline Humphrey's observation that, in Mongolia, "landscapes are more in the nature of practices designed to have results: it is not contemplation of the land (gazar) that is important but interaction with it, as something with energies far greater than the human" (1995, 135). Anthropologist Morten A. Pedersen, who has conducted field research with the Mongolian-speaking Darhads and the Tyvan-speaking Dukha of northern Mongolia, also emphasizes interaction with the landscape, in which "meaning is 'drawn' or 'elicited' from the landscape via a continual engagement with it in the form of both everyday and more ritualized nomadic practices" (2016, 222).

I conducted the field research for this study between 2015 and 2018 in partnership with ethnomusicologist Robert O. Beahrs[8] in Tes-Khem Province, Tyva. In our work we collaborated with my home community in Ak-Erik. I reflect on challenges of being an Indigenous researcher and working in my home region for my doctoral dissertation. Together with my grandparents, we shared camp sites and jointly herded livestock with some of the herders

Figure 3.1 A Kodan of the Aal with Two Families' Yurts. South Tyva, Summer 2018. *Source*: V. Peemot.

of my patrilineal clan.[9] While most of the herders see in me kin and akaa I, as a woman, find myself in an (involved) outsider's position when discussing horse issues, because Tyvans generally consider the horse theme a man's domain. However, I noticed that a researcher's position can be gender-neutral.[10] During our stay, we worked with herders in their seasonal camping sites in various locations: in the Agar steppe in winter, and in summer in the arid area around Lake Shara-Nuur, as well as in the boreal forest of the Tangdy-Uula Mountains.

THE *AAL* AS A SOCIOECOLOGICAL UNIT

Herding communities in South Tyva understand aal as a human-animal unit characterized by geographical inconsistency. An aal can change sites within a seasonal pasture; its membership is temporal with the number of family yurts fluctuating, while the exchange of food and working skills is ongoing. Field research in South Tyva suggests three main types of aal: the single-family aal with its single yurt, the aal consisting of three yurts belonging to two nuclear families (including their parents and a married son, see Figure 3.1), and the multi yurt aal of two or more families, which may or may not have kinship ties. When we visited the Ak-Erik herders in their summer pasture at Lake Shara-Nuur in

2016, there were thirteen yurts composing nine aals. Two of them consisted of two yurts each, and they were similar in their social structure. The youngest married son shared the campsite with his parents. This type of aal fits the description of "the stem family" among Buriats before collectivization, "a residential unit consisting of the married heir living with his parents" (Humphrey 1983, 271). Five aals consisted of a single-family yurt each. Four other families had divided into two aals, each making up a unit of two families consisting of distant kin. The following winter, the families moved to their single-family winter campsites [Tyv. *kyshtag*], with the exception of two stem families. In their case, each nuclear family owned a separate house in a shared winter site.

Though the contemporary aal structure is not defined on economic grounds[11] but on kinship ties, one can observe intra-aal economic relations in the way of natural exchanges of food and skills. The families of an aal are involved in a mandatory food exchange—part of the day's main meal is given to neighbors. When a goat or sheep is slaughtered for meat, the neighbors receive a set of cooked intestines and broth in addition to one limb with adjacent joints, the latter of which may be a front limb with scapula and ribs. Later the neighbors will return the favor with a matching amount of meat. Anthropologists, Matthew J. O'Brien and Todd A. Surovell, recently observed the same custom among Dukhas, the Tyvan-speaking reindeer herders in northern Mongolia:

> In the span of 25 observation days, we were present for meat sharing of a domesticated reindeer and goat. In both cases, the primary owner of the carcass distributed equal portions of internal organs and a limited amount of meat-bearing bones with the other families in Camp 1. Given that Camp 1 is mostly linked by primary linkages, this pattern of sharing is expected. (2017, 117)

The intra-aal exchange of skills includes joint herding of livestock, help with making traditional clothing, and conducting seasonal livestock related work, including shearing sheep, castrating, and weaning livestock. A single-family aal takes care of its own animals. All livestock of a parents-son aal is herded and milked jointly. The owner makes decisions about selling animals or slaughtering animals for food or sale at the meat market. While nuclear families have separate budgets, they share expenses for conducting seasonal work and buying vaccines, medication, and forage for animals. This also includes repair materials for a winter place and gasoline for transhumance. In the type of aal that unites two or more families belonging to an extended family, livestock is taken care of jointly in summer pasture. Usually sheep and goats are combined into one flock, being kept in the same corral at night, and herded in turns by families. Cows can be herded and taken care of separately or jointly by both families. Herding families also care for the livestock of relatives who live in a village or town. "Urban-rural exchanges are part of the social networks that have helped sustain both herders and city folk during the economic transition period," writes Fernández-Giménez in her study of post-socialist Mongolia (2002, 64–65). The herding family and absentee herders (ibid.) are connected by obligations: the latter help with seasonal works and during transhumance, and they will host the children of the herding family during the school year at their village or town home.

In summer 2018, we stayed with Soian Nogaan-ool and Ailangmaa in their summer encampment by the Khaialyg-Semis-Tei spur of the Tangdy-Uula Mountains. The second yurt in the aal belonged to Dizhitmaa Shoraan and Aiana, prominent herders in their early forties from Bert-Dag Sumu. One day Nogaan-ool, along with his son Arapchai, went to cut firewood at the edge of the forest. No neighbors offered help with woodcutting or transporting. The next two days Dizhitmaa Aiana sheared the Soians' sheep. This reflects a customary division of labor to the work of the yurt, work of the kodan, and works that do not belong to the intimate domestic sphere of the family. Works of the yurt are cooking, cleaning, and washing. Works of the kodan are outdoor tasks, performed around the yurt: cleaning the territory and animal corrals from dung, keeping the kodan tidy, milking and accommodating animals in shelters, and watering and feeding hay to stock. No help is expected in either yurt or kodan labor, except from close kin. The making of clothing and livestock harnesses, as well as major seasonal work, such as shearing sheep, trimming manes and tails of horse yearlings, and castrating livestock are considered above what can be expected of a single household, and help of neighbors or visitors is welcome for these tasks.

A multispecies community aal inhabits tame landscapes—*kodan* and *khonash*. Their meanings are distinct, and this distinction resides at the core of herders' interactions with the landscapes they live in. It also reflects and affects their relationships with horses.

Livestock of the Land

The Ak-Erik herdsmen emplace their horses differently from other domesticated species because horses are defined by their not-quiteness.[12] On the one hand, a horse is understood as *cher maly* [Tyv.], which translates as "livestock of the land" (or wilderness). By contrast, cows, sheep, and goats are *kodan maly* [Tyv.], "livestock of the *kodan* campsite territory." A horseman of the municipal unitary enterprise "Choduraa" from the municipality Beldir-Aryg with pastures neighboring Ak-Erik explains:

> It is livestock of the land; it does not belong to livestock of the encampment. It spends nights out in the land. It is not suited for fences. The same thing is said about yaks: they suffer inside fences, lose weight, and become barren. Cows,

sheep, and goats are livestock of the encampment. (Melestei Oiun, Fieldwork interview, 2015)

As livestock of the land, horses have greater autonomy than other domesticates.[13] They graze on distant pastures, often unattended. Herders estimate approximately where the herd may be and check on them at the pasture. The horse is also understood as a mediator[14] between humans and *Tangdy* eezi, the nonhuman master of Tangdy, who holds the highest position in the human-nonhuman hierarchy. On the other hand, horses are close to humans. Consider how two equestrians, father and son from Ak-Erik, related to their horses:

> You know the habits (of your horse) just like you know your wife's character at home. You feel that she is angry with you and you try to avoid conflict. In the same way, you know your horse. When another person tries to catch your horse, it would not let it happen. . . . The horse recognizes his human. For instance, the horse will not leave that human anywhere. A human and horse grow close to each other, they become inseparable. (Alexei Byzaakai, Fieldwork interview 2015)

Alexei's father, Andrei Byzaakai, explained, "Riding a horse is a pride to pastoralist people, to both men and women. They ride horseback and enjoy their air and nature, drive their livestock, sing, and do *khöömei*,[15] laugh, and feel happy. It is that—freedom in life" (Fieldwork interview 2015). The horse's simultaneous closeness to wilderness and to humans emplaces horses differently than other nonhumans in the multispecies landscapes, which I will discuss in this chapter.

A GENDERED APPROACH TO HORSE HUSBANDRY

The Ak-Erik herders, who change seasonal pastures four to five times per year, continue to use a yurt [Tyv. *ög*],[16] which is easy to disassemble and transport by truck. Only a winter campsite is usually equipped with a modest-size log house. The yurt's shape, its components and fuel, as well the choice of its insulation materials are the result of mobile communities' adaptation to climate, landscapes, and the pastoralist economy. Sheep felt [Tyv. *kidis*] covers the yurt's top [Tyv. *deeviir*] and floor [Tyv. *shirtek*], and the structure is held tight by horsehair belts [Tyv. *kurzhaangy*] and camel hide 'nails' [Tyv. *kökteshki*],[17] the latter of which are still preferred for connecting wall panels. Each of these materials is chosen for their respective qualities. For instance, horsehair is resistant to humidity; when wet it does not stretch or loosen. The yurt's wooden lattice wall panels [Tyv. *khana*] and rafters [Tyv. *ynaa*] are traditionally made of lightweight willow. The circular crown [Tyv. *toona*] is made of birch.

During our stay at Ak-Erik, the entrances of the yurts all faced south or southeast.[18] The interior space of these yurts was referred to as western or eastern sides, the latter being the "female" side. It hosted a kitchen area, the woman's bed, and, next to it, a section [Tyv. *üzük*] to store food supplies. Food related work (cooking, separating butter from milk, washing dishes) was done on the "female" side. Guests were not allowed here, except women who were close kin. A visitor, upon entering the yurt, would move clockwise, that is to the western side, which is the "male" side. Here the visitor would sit in front of the man's bed. Horse equipment for everyday use was kept on the western side. Saddles were kept in a storage area adjoining the bed from the entrance side. Bridles, halters, lasso-*sydym*,[19] hobbles, and whips were hung on a special hanger [Tyv. *chagy*]. It was made of a debarked and smoothened tree trunk, approximately 140 centimeters long and 10 centimeters in diameter. Small branches left on the upper part of the trunk served as hangers. The chagy was placed at either the head or footboard of the man's bed. Horsehair rope weaving and processing of hides and leather for traditional clothing and equipment was done in the yurt's "male" side. Although the yurt is primarily a symbolic place for human-animal encounters, actual animals sometimes are sheltered in the yurt. If weather conditions require, lambs, kids, and calves may stay inside a small indoor paddock, which is set up in the western section, close to the entrance. Because of the herders' understanding of equine autonomy, foals are never sheltered in a human dwelling.

The distribution of horse-related items, and the constructing of horse equipment inside the western part of the yurt, reflected the gendered approach to horse husbandry among Tyvans. The horse herd, as a whole, related only to its male owner, although all livestock was seen as family property, and individual horses did belong to individual persons of either gender. In contrast, cattle and sheep were mentioned in relation to either male or female owners. The gender inequality in horse husbandry was further evident in the custom of bestowing livestock upon children. While girls and boys received sheep, goats, and cows in equal measure, boys tended to acquire more horses than their sisters did. While looking for horses in the Tangdy-Uula Mountains with Arapchai, the fifteen-year-old son of Soyan Ailangmaa and Nogaan-ool, we asked the boy how many horses he owned. He explained in the Tyvan manner avoiding actual numbers: "When I was just born, my father gave me a horse. Since that time, I have received horses as gifts. Now I own one stallion band. I have horses in other stallions' bands too" (Fieldwork conversation 2018). The family owns three stallion-bands; all together, there are approximately

120 animals. Two-thirds of them belong to Arapchai. His parents and two sisters hold the final third.

INSIDE THE YURT: HIERARCHY OF THE HORSE

While customs demanded respect for horse tack[20] in general, some horse-related objects were of especially high symbolic value. Their high social regard was evident in how and where they were located in the place of honor [Tyv. *dör*]. The place of honor was found on the northern side of the yurt, opposite the entrance, and behind the hearth. Guests entered the dör only by invitation.[21] Due to seasonal transhumance, all the items found inside a yurt were chosen carefully. Every item in the mobile household had to be assessed in terms of its ability to be transported. Under such conditions, the presence of horse-related objects in the yurt was not accidental. To provide an example, I shall analyze the horse-related objects inside the dör of Aldyn-Kherel Dörbet-ool's yurt. In the center of the northern part were medals won by Dörbet-ool's racehorses. The medals were hung along with sacred scarves [Tyv. *khadak*] of various colors. The northwest of the dör held Tyvan saddles and bridles, as well as Dörbet-ool's traditional set of knives, pipe, and flint (both of which hung on a belt when in use). I took a photo of these traditional items in Dörbet-ool's yurt in 2018 (see Figure 3.2).

Previously, in 2013, I had seen identical items in the yurt of Dörbet-ool's parents. His father, Roman Aldyn-Kherel, had shared with me stories of some of the items, including where they had been made. He also had explained how the leather parts of bridles and halters were repaired every few years. Roman akaa had told me on that visit that the ownership of this equipment, including the livestock, would transition to his youngest son Dörbet-ool once he would get married and start his own family. In 2015, when Dörbet-ool got married, he did indeed inherit both. In addition to the horse-related valuables, the dör area also accommodated, in different variations, the Buddhist altar, various family photographs, and official letters of appreciation for herding achievements, as well as equine statues and pictures. Though the family's history remained undocumented in writing, it echoed from the stories that came with valued items, especially the horse tack that passed from father to son. For this reason, it is fitting to approach the place of honor as the locus of a family's memories. Correlations between an object's value, its placement in the yurt, and the social connections the object carries have also been observed among Mongols: "present-day Mongols persistently categorize objects in terms of their position in space" (Humphrey 1974, 13).

Figure 3.2 Roman and Dörbet-ool Aldyn-Khere in their Yurt, Ak-Erik. *Source*: V. Peemot.

By emplacing the horse in the hierarchical and gendered spaces of a yurt, the residents revealed their recognition of the horse as a being of high symbolic standing. In the yurt, the horse found reference in the most honored northern part, where it was attributed primary connections to men in the western "male" side of the structure. By extension, the northern part of the yurt served as locus for the memory of the aal as a multispecies socioecological unit.

THE KODAN AS LANDSCAPE OF MULTISPECIES INTERACTIONS

This chapter aims to reveal the striking difference between the dominance of the horse in the inner space of the yurt and, by contrast, its absence from the territory surrounding the yurt. While the former emphasized the horse's symbolic value, the latter revealed the autonomy of equines. People used the Tyvan term *kodan* to refer to the localized landscape, the area which hosted the multispecies social unit of the aal. Here herder-livestock and livestock-family or aal relations could be observed best.[22] Our fieldwork with South Tyvans suggests that the aal and kodan in Tyvan do not aggregate conceptually, although they can be used together as a cumulative term. The kodan included a yurt, or a group of yurts, corrals, and shelters for livestock, a tethering post for horses [Tyv. *baglaash*], and an adjacent territory which remained unmarked. In spite of the absence of markings, humans and nonhumans were ever sensitive to its invisible boundaries. "In-habiting" the land, in Humphrey's sense, is "interaction with it" (1995, 135). Inhabiting the kodan required from humans[23] and nonhumans that they follow certain rules of interaction, which varied, between "livestock of the kodan" and "livestock of the land." I use the architecture of the kodan and the amount of time each species spends in the kodan as analytical tools to better understand the autonomy of horses. A comparison with other nonhumans' interactions with the kodan further helps clarify the distinctiveness of the horse-kodan relationship.

The nonhumans' perception of the kodan and their sense of belonging to the aal (and to the kodan) were best observed in dogs' behavior. Most dogs seemed aggressive toward strangers, but they would come to accept the temporal belonging of guests to the aal after the guests had spent some time in a yurt, and especially when they had seen people of the aal treating the visitors in a friendly manner. Dogs also distinguished livestock of their own aal from that of other aals'. "Unrelated" animals were chased off the kodan. People distinguished Tyvan dogs based on work tasks in which they excelled. Thus there was the hunting dog [Tyv. *angchy yt*], the herding dog [Tyv. *kadarchy yt*], and the kodan guard dog [Tyv. *kodanchy yt*].

> [Speaking of his own dogs]: These dogs are Tyvan dogs: pure, not of mixed breed. They are old, though. They are very good in the kodan, especially in guarding against wolves. One of the dogs will jump at a wolf, but winning a fight is hard for him. The other dog is scared; he will just circle around a wolf at a distance. They can deal with one wolf, but they cannot stand against a pack, or surely, they will be eaten. I thought the dog would long have been eaten by a wolf, but he still lives, the poor thing. Now he is ten years old. When a dog reaches ten years, he becomes heavy, my daughter. He does not move much, a bit lazy, sleeps, and stays close to the yurt and guards only the *ödek*.[24] (Oleg Sambuu, Fieldwork interview 2017)

The other nonhumans' ways of interaction with the kodan further reflected their being distinguished either as livestock of the campsite, or as livestock of the land. The majority of livestock—sheep, goats, and cows—were bound to the kodan. In Tyvan, the term kodan is also part of the expressions "a flock of sheep" [Tyv. *kodan khoi*] and "a herd of cattle" [Tyv. *kodan inek*] (Tatarintsev 2004, 168). Livestock of the kodan were prominent in the architecture of encampments. At a summer campsite, one could find open corrals for flocks of sheep and goats. Most herders in the area had started to erect summer corrals for their cows by the mid-1990s, when livestock theft had become epidemic. Oxen and young females, who tended to wander off during the night, were driven into corrals for the night, while cows, whose calves were corralled in another enclosure, would stay on their own within the campsite's territory. Herders who spent their summers in the boreal forests of the Tangdy-Uula Mountains would relocate their corrals whenever heavy rain, combined with the trampling of animals, had caused the ground to turn into mud. If need be, a family would relocate their entire campsite, along with its yurt and corrals, to nearby untampered grounds once or twice per summer.

In winter, the herders of Ak-Erik would turn their livestock enclosures into what anthropologists David G. Anderson et al. call "structures of comfort" (2017, 406). This was a challenging task given the climatic conditions. Their winter grounds were located in the steppe between Agar Mountain and the River Tes. Only the riverside had some trees, among which were poplar, willow, and bushes of sea buckthorn berries. The rest was an open area where winter temperatures were generally lower than in forested places. "It is minus 53 [Celsius] here. I have just come in after cleaning the cow corral. It is cold," reported my cousin Alina when I called her in late January 2018 to ask how the family was faring in this cold weather. News about the harsh conditions in Tyva had made headlines in the Russian media after two people in neighboring Erzin Province, also a steppe zone, had frozen to death after their tractor ran out of fuel. In the Tyvan type of pastoralism, livestock graze on open pasture every day of the year. Even if it was minus 53 degrees Celsius, the herders would still drive their animals to pasture, doing their everyday chores, cleaning corrals and kodan of dung, feeding the weaklings and, if calves had been born, taking them out to nurse, and milking the cows.

Khorlai Langaa emphasized the importance of keeping the kodan territory neat. "Inside of the kodan—this is the kodan of the summer campsite, and there are kodans in autumn, spring and winter sites—one has to regularly clean out the livestock's urine and dung and keep the kodan clean" (Fieldwork interview 2015). On especially cold winter days, thick haze blanketed the Agar steppe. Knife-sharp air would pierce our lungs with each breath. Under these conditions, the herders' care for their animals was an example of how "human agents will go to great extents to make enclosures comfortable and attractive to individual animals" (Anderson et al. 2017, 407).

The herders moved into their winter encampments at the end of November, and they began to shelter their animals overnight around mid-December. The architecture of a winter kodan included a small log cabin for the herders, a corralled hay supply for their animals, and various livestock enclosures. Warm stables for sheep and cattle had been built side by side, to benefit from a shared wall. Yet each stable differed in height and size. The insulated shelters used for sheep and goats were capable of holding a few hundred animals. Horizontal logs with a diameter of 15–20 centimeters supported their roofs, and their log walls were caulked with moss and cow dung. In November and December, people gathered fresh dung in the mornings, which they applied to the moss in the walls' seams. Once the dung had dried, it formed a smooth surface. The roof structure itself, which was angled in a slight A-frame, consisted of thin, tightly spaced, unhewn tree trunks, which were covered by straw or reed. Dried dung was used to insulate and seal the top. There were no windows in the stable's walls. Depending on a structure's size, two to three open skylights provided natural light and ventilation. These skylights were left open overnight, while rugs or wooden planks were used to seal them off during the day. Skylights were the weak point in this architecture, as they enabled wolves to enter. The angled roof made for uneven ceiling height inside the structure. An adult could stand up straight only at the center of the building, and people had to duck on entering. At the center of each stable, a corral with two or three sections held the lambs. For their safety, newborns always stayed within these areas. Outside the insulated barn, the connected *kodan kazhaazy* [Tyv.] provided a small fenced open-air area. It formed part of the encampment's kodan, and the lambs would spend daytime here. The kodan kazhaazy also served as a hay feeding and watering area for lambs.[25]

Herders would consider the age, gender, and physical condition of their cattle when deciding whether to keep them in a warm barn or in an outdoor structure during winter nights. Adjacent to an insulated sheep barn, the herders had built a warm cattle barn, resembling the sheep barn structure, but with a number of differences. A higher ceiling, a full-size entrance, and a smaller footprint made the structure a suitable shelter to a few dozen cows. Cows, heifers, and one to two-year-old male cattle spent their nights inside this warm shelter. Inside, adult cows were tethered to the walls, with ropes long enough to lie down, and short enough to prevent fighting. The herders also kept a distance between individuals they knew had a dislike for each other. Adult oxen and the herd's reproductive bull would spend their nights outdoors, sheltered under a roof structure partially covering a corral. The partial roof structure was walled in on three sides (one of them being the wall of an adjacent barn), to provide wind protection. A thick layer of dried dung was spread regularly to insulate the stable floors against the cold.

Calves were born between January and April, with a peak in late February and the beginning of March. The herders would bring calves born in the depth of winter into their cabin for a couple of weeks. Here the calves would stay in a wooden box with dry cow dung bedding. Beginning in February, the calves spent their nights in a separate warm barn. This was a relatively small building, sheltering twenty to thirty calves. Dried cow dung was preferred over sheep dung as bedding in the calves' barn. The herders assumed that sheep dung[26] had more fine dust-like particles, which the calves might inhale, damaging their lungs.

Cows, sheep, and goats would casually spend time in the kodan during all seasons of the year. In the heat of summer, sheep would take their daily naps in the kodan, and calves would run in and out of the kodan all day long. Even in winter, cows, sheep, and goats spent much of their time in the kodan. The herders would leave the young lambs and calves inside the corrals, where they fed them additional forage. Pregnant cows and ewes spent the final days prior to birthing inside the kodan, or at least nearby, as they would often require assistance during delivery. The newborn calves, including lambs born during cold winter months, were then taken into the herders' cabin or yurt.

THE KODAN AND EQUINE AUTONOMY

Ak-Erik herders engaged a series of different "architectures of relationships" (Anderson et al. 2017, 398) with their animals. Cows, sheep, and goats who spent more time near the yurt, and who were regularly sheltered in protective enclosures, clearly possessed less autonomy. Conversely, animals, such as horses, which made significantly less use of the architectural affordances of the kodan, were exercising a much higher degree of autonomy.

There are only two items that point to the presence of horses in the kodan—the baglaash tethering post and the open enclosure. The tethering post for horses is symbolic of Tyvan mobile pastoralism and its hereditary land ownership, which is defined by customary law. Tyvan author Kuular Shomaadyr, in his novel "Baglaash," presents the tethering post as a sentient witness to the life of a herder, who feels the warmth of a woman's touch and deeply cares about his human family. Baglaash, in the author's description, looks like a Tyvan man—his features carry signs of hardship and weather.

> The Baglaash has stood in place for such a long time that its face has paled; its proud head and trunk have cracked from the merciless winds and storms, snow and rain, heat and cold. If you look carefully, you will be reminded of the wrinkled face of a man whose life has been full of hardship. Especially when you look in the evening twilight, against a light skyline, you can make out an older man, wearing a high Tyvan hat. It has not regretted leaving the forest and helping the human; it has even been proud of it. In the forest, who would see it? [Here] it ages and sees time running by—all the while standing in one place! Ekh! Over the course of time, men, girls, youths and the old have come to it! Greeting and wishing farewell to them all, it hardly noticed time passing as it aged. What now? Has it been left at the old campsite? No! My time may have passed, but a new time has come. Time does not stand in the same place! (Kuular 2002, 3–4)

Until recently, the baglaash was the only reminder of the horse in the kodan. Well acclimated to harsh winters and self-provisioning, equines require no shelter to protect themselves from weather and wolves, nor do they require human assistance in finding watering sources. They also require no assistance during the foaling period. Horse enclosures are an innovation of the mid-1990s, when a wave of horse theft hit the area. This has changed the way horses were being herded during our fieldwork. Most herders would round up their horses in a corral every night instead of letting them range freely. Still, the herders ensured a group of horses spent minimal time inside a corral, which usually was from 10:00 p.m. to 5:00 a.m. in winter, and from midnight to 4:00 a.m. in summer.

Horse enclosures were made of unhewn tree trunks, and they provided no shelter from weather. Their sole purpose was to protect against thieves. In early spring, the corral was also used in trimming manes and tails, and for branding and castrating. In winter the horses might linger a bit longer in the kodan because of the salt that would be laid out for them. In other seasons, Ak-Erik herds satisfied their salt requirements by licking water and salt deposits on Lake Shara-Nuur.

The herders with whom I worked on a regular basis only rode about ten to fifteen horses from their herds. They did not train the rest of their horses for riding.[27] The untrained majority remained unaccustomed to human proximity, although herders did interact with untrained mares when milking them. However, the only herder in Tyva whom I have seen milking mares regularly was Oiun Melestei from the enterprise "Choduraa." At his summer encampment, where the milking would take place, Melestei had built a narrow paddock for milking mares. In 2015 there were about thirty mares to milk, and none of them, according Melestei, had ever been trained for riding. In this case, the structural specifics of the paddock were reflective of the mares' strong sense of autonomy. The untrained mares roamed freely for most of their lives and, as a result, were unaccustomed to human touch. Melestei placed them in the narrow enclosure of the paddock to restrict their movement. The paddock enabled milking, while it protected the milking person from being injured.

During our fieldwork with Altai Tyvans of Western Mongolia, my research collaborator, Robert, and I observed a different situation with mares that were accustomed to regular milking. Here only first-time mothers had to be restricted. Using a lasso, a man would keep one of the horse's legs raised, while a woman would milk the animal without the risk of being kicked. This was a simplified method of restriction, which, according to the herders, had to be used only during the first few days of milking. Over the following days, mares became accustomed to the milking process.

Equine dairying in the Altai Mountains was distinct in several ways from other livestock dairying. We observed, for instance, that people always milked yak and sheep inside the kodan, while mares were milked outside of it. A few hundred meters away from the yurts, the men of the household would erect a *chele* [Tyv.]. The chele is a traditional tether for foals, which consists of two wooden posts with an argamchy lasso stretched between them. The herders would use the chele during the milking period for approximately two months every year. During this time, they would tether foals during the day, and release them in the evening to spend the night with their mothers.[28]

Although Ak-Erik herders trusted their horses' self-sufficiency, they had adjusted their mating period to ensure foals were born after the coldest period of the season. Mares of the Tyvan breed require no assistance during delivery. They give birth at pasture, shortly after which their foals get up to stand, nurse, and follow the herd. Unlike newborns of other livestock, the herders would never bring a foal into a human dwelling. As a herder noted:

> It is good to release a stallion to his mares for mating a bit later, so mares will deliver foals from the end of March to the end of May. Now it is not so easy because there are untaught and bad-tempered stallions. Stallions of our time are not like the old time Tyvan horses. Tyvan stallions would not leave a mare barren despite of their numerous bands. Now these half-blood Russian stallions come to maturity at a later age. Even chosen as a stallion, they walk idly until they are four or five years of age. A Tyvan stallion already starts to gather his own mares and band at two years of age. In fact, it is useful to keep a record on days [of mating]. If you see today that a stallion is mating with a mare, you write the date down. A mare is pregnant for ten months and delivers a foal just at the beginning of the 11th month. So you know the delivery date and you can put that mare in a corral. In the old times, people kept this kind of record. They would know dates even for sheep and cows. The mother's udder starts to fill when the child inside her starts to grow a fluff. A mare does not need assistance [during delivery]. (Aldyn-Kherel Roman, Fieldwork interview 2015)

THE KHONASH: KEEPING HUMAN-NONHUMAN STORIES

For mobile pastoralists of South Tyva, humans and nonhumans, including animals and masters of the land, constitute a shared local life-world, best described as "co-existence." The identity of an Ak-Erik herdsman emerges from continuous "mutual interrelations" (Anderson 2000, 116) with nonhumans and landscapes. Khonash is a term that encompasses the seasonal camping site of a multispecies aal, the immediate territory of the kodan, as well as the surrounding water sources and pastures. The khonash mirrors the cyclical movement of a herder's life, as the aal returns to its khonash every year in the same season. As the season drew near, the herders and animals of our fieldwork would begin to long for their khonash. They looked forward to the affordances found in the landscape of the khonash, which made it attractive in the given season. Animals would express their longing for the khonash by attempting to move in its direction. Horses usually initiated transhumance toward the next khonash. At the cusp of a new season, it became more laborious for herders to keep their pressing herds in one place.

A herding family moves between several khonash during the year. The only permanent khonash is the winter site, which is occupied from late November to mid-April. The herders invest in its architecture disproportionately, building a cabin for themselves and insulated enclosures for their animals. After winter in the Agar area, Ak-Erik herders moved to two different climate zones. Some families spent the rest of spring by the River Tes and continued their transhumance to the boreal forest of the Tangdy-Uula Mountains. Conversely, thirteen to fifteen families would move to the arid steppe of Shara-Nuur from April to November.[29]

Spring, summer, and autumn khonash all blend into the wider landscape. Each transhumant aal re-constructs its kodan, as it sets up its yurt and proceeds to build light enclosures for its animals. The kodan with its temporary structures is only a part of the seasonal khonash, which also includes water sources and pastures. Various parts belonging to the khonash are not immediately adjacent. Instead, a network of habitual paths leads to them. Among such places are water sources, used jointly by humans and domestic and wild animals. In contrast to the kodan, which is always "attached" to its aal, the khonash has a name of its own. The story of the shared lives of humans and horses is "written" into the landscapes of the khonash, as several generations of one family will use the same sites.

Customary law regulates hereditary khonash ownership. I was able to witness the symbolic transition of livestock and khonash ownership from father to son at Samba Langaa's wedding, which took place in the summer at Kara-Chyraa khonash in Shara-Nuur in 2013. Here Samba's father, Khorlai Langaa, announced the transition of "inherited land" to his youngest son:

> We have inherited this land from our ancestors. Shara-Nuur is a rare place in Tyva. It has spacious pastures with hot-grass. There is a salt lake with healing waters and mud, small fresh-water lakes and Naryn River with its clear water. Five of our children live in the village and town. Everybody has a family and a job. I have told my youngest son, 'We, your parents, are not young. You will stay with us and take care of our livestock and camp sites. (Transcribed from author's video recording, 2013)

This transition speech was followed by a precious wedding gift from Khorlai akaa to his son—a four-year-old Tyvan breed bay gelding, a promising racehorse. Khorlai akaa explained that the bay horse was descended from a special stallion he had received as a gift at his own wedding from his father Dargan Langaa thirty-five years earlier, also at Shara-Nuur. Samba's new bay gelding had the benefit of being an embodiment of entangled human and horse lineages; both belonged to the same landscapes. Although Khorlai akaa explicitly said that he was giving his livestock and khonash to his son Samba, in fact, the actual transition would be a continuous process. Samba had been prepared to "take care of livestock and campsites" for many years:

> Since I can remember, my father's parents have been around. They spent summers by Lake Shara-Nuur in the Agar. My parents would go to Naryn River for the summers. Prior to attending school, I followed my father as he went to round up the horse herd. First, I sat behind my father on the back of his horse. Later, I learned to ride by myself. I went to the school in Ak-Erik. Since completing the ninth grade, I have helped my parents and stayed with the livestock all the time. Then I entered the army, where I spent six months in Khabarovsk and a year and a half in the Far East. Within a year or so of returning from the army, I got married. Her name is Dolaana, and she is from Torgalyg in Ulug-Khem. Because I grew up here from an early age, I continue to follow my parents and keep livestock. My elder son's name is Baldan-Seren, and our little one is Choigan. One of them is two and a half years old; the other is just six months old. I will raise one of them to be with livestock. After me, he will connect his life with livestock. (Personal communication, July 2013)

Some researchers have described the life-world of mobile pastoralists as "hybrid triadic associations of humans, animals, and the landscape, which construct a specific socioecological niche by their mutual interaction" (Stépanoff 2017, 381). Anderson (2000) has called this relationship a "sentient ecology," in which each constituent possesses the agency required to engage in a productive conversation with the others. The multispecies aal of Ak-Erik herders is a good example of such an ecology. Here the herdsmen bond with their horses differently than they do with other livestock. The horse, being understood as "not-quite-livestock" (i.e., close to both wilderness and humans), naturally forms a communicative bridge between the kodan and the open steppe, which stretches beyond the domain of the kodan. At the same time, the horse dominates the symbolic space of the yurt, affirming its high value to the aal. Meanwhile, its relative physical absence from the kodan further supports the notion of equine autonomy. Finally, the landscapes of the seasonal khonash encampments serve as keepers of stories and memories shared by different generations of the multispecies aal.

NOTES

1. I am grateful for the genuine support of all Ak-Erik herders. Here I name only a few of them: Roman Aldyn-Kherel, Andrei Byzaakai, Seren-Shimet, Sergei Kaldang, Nogaan-ool, Ailangmaa Soian, and Khorlai Langaa († 2018). Special thanks go

to my cousin Aisula Erendei, whose family hosted me at Ak-Erik, and to akaa Vladimir Orus-ool for his practical help and thoughtful insights into the world of Ak-Erik equestrians. I am grateful to my research partner Robbie Beahrs for his collaboration, friendship, and feedback on this chapter.

2. This and all further translations from Tyvan to English are made by the author.

3. More on the word *akaa* in Bavuu-Surun (2018, 210).

4. More on the ethnonym *Soian* in Tatarintsev (2015, 341–352).

5. This is the plural form of the noun *akaa*.

6. The group of *hot*-grasses [Tyv. *izig o't*] includes plants of several genera: Kochia, Artemisia, Ephedra, and Parmelia (Arakchaa 2015, 79, 105–110).

7. My use of the concept of emplacement in human-animal relationships is inspired by the work of anthropologist David G. Anderson (Anderson 2000; Anderson et al. 2017). In my view, emplacement best describes an actor's positioning within a shared multispecies landscape. It refers to a scenario in which humans and nonhumans engage each other. In the context of my research with equestrians of South Tyva, emplacement includes wild and domesticated animals, as well as nonhuman and non-animal beings inhabiting certain landscapes.

8. Currently a lecturer at the Centre for Advanced Research in Music (MIAM), Istanbul Technical University.

9. Nonhumans also participate in this network of connections. In spring 2017, we visited Sergei Kaldang at his winter place. My grandfather, Kadyp-ool Soian, and Sergei's father, Kaldang Soian, were siblings. When I was eleven, Kaldang Soian gave me my first horse—a red roan mare.

10. In spring 2017, I was invited by several equestrians to participate in their seasonal work of castrating and trimming tails and manes of two-year-old horses. Our group consisted of nine men—my relatives and research collaborator Robert—and myself. During the day, I sometimes felt excluded from the company. For instance, as a woman, I was excluded from participating in the drinking of strong *chifir* tea, which was a male activity. As the only woman in my unmarried cousin's encampment, I was also relegated to cooking for everyone.

11. For more information on the economics of aal communities in Tyva, see Potapov 1969, 105–144 and Vainshtein 1980, 243. Bat-Ochir (1996, 71) also writes about *khot-ail unions* of Mongolia.

12. See Peemot (2017) for more information on the *not-quiteness* of horses.

13. Similarly, reindeer herders in northeastern Tozhu Province respect the autonomy of their reindeer (e.g., Stépanoff 2012). Tozhus often refer to reindeer as "wild animal-like" (Donahoe 2012, 111). Thus, keeping reindeer in a corral is seen as unreasonable. Hunter-herder Buyan Ochur (personal communication, June 2014) from the eastern province of Tere-Khöl told me about a *badakhshan* (cf. Pedersen)—a human-animal [Tyv. *kizhi-ang*]—a meter long tall being with brown fur and round eyes, who lives in the mountainous taiga. Badakhshan dislikes it when people cut a tree "with no good reason." Once, the account goes, herders in Tere-Khöl built a corral and enclosed their reindeer for a night. By morning neither reindeer nor corral could be found. Later the herders found the trunks they had used in the corral. Someone had carried them back to the forest, where they lay by their stumps.

14. A Tyva herder Chimed Dondog who lives in the Altai Mountains of Western Mongolia explained to me in 2016: "If the shrine-horse is well taken care of, our Altai-sky is happy. If treatment of the horse is not good enough, our Altai-sky is discontented and offended."

15. "Commonly called 'throat-singing' in English, xöömei refers to solo-voice drone singing practices associated with nomadic pastoralism, male guttural virtuosity, and multiphonic overtone melodies with historic roots in the Sayan-Altai Mountains of Inner Asia" (Beahrs 2017, 61; see also Beahrs 2019).

16. It is not the only dwelling of herders; most families have houses in a village or the republic's capital, Kyzyl.

17. Holes are made where lattice wall sections overlap; a short piece of soft camel hide is run through two wooden strips to connect them. The dried camel hide hardens and becomes water-resistant.

18. As an exception, some yurts in the encampments of the Tangdy-Uula Mountains are oriented to the east. Similarly, cardinal directions apply to the horse's body: "The brandmark is placed on the western hind leg," explained Altai-Tyvan herder Chimed Dondog from the Adai-Irgit clan in Western Mongolia. I observed that Dondog's horses had their branding on the "western" side of their bodies when their head is oriented to the south.

19. Ak-Erik herders manufacture lassos [Tyv. *sydym*] from thick cow or ox hide. They use various techniques to process and cut the leather, combined with processed sheep and goat horns, which function as stoppers. The phrase, "he's a man who does not drop his sydym," refers to a person able to catch a horse with a lasso-sydym on first try.

20. It is customary to keep the horse tack neat and clean. The tack must not be left on the floor, and it is forbidden, especially for women, to step over the equipment. The high regard for horse tack may also have to do with its monetary value—decorated with silver and handmade by master crafters; the traditional saddle and bridle are valuable items.

21. Working with herders across different provinces of Tyva, I noticed that, while "invitation-only" is a strict rule, the gender distinction is blurred in the dör. However, I have experienced a gender difference in Mongolia. In the yurt of Tyvans in Tsengel village of Western Mongolia, the hosts invited me to the dör. A Tyvan man from the Mongolian Khvod province came in after me, refusing to sit "below a woman." The hosts reasoned that they saw a respected guest in me, as I was "kin from Tangdy-Tyva."

22. A phrase exemplifying this would be, "The cows of the neighboring aal have returned to their kodan."

23. One of the first rules I learned in riding horses, was how to act on horseback inside the kodan. A rider must move in and out of the kodan at a walking gait, accelerating only beyond kodan boundaries.

24. The word *ödek* in the quote is used as a synonym for kodan, the territory within a campsite. Literally, ödek stands for the dry livestock dung, which is used as insulation material in the dirt floors of winter animal shelters.

25. Insulated shelter architecture has undergone some historical changes over the years. During the Socialist period, herders from Choduraa Sovkhoz (neighbors of Ak-Erik), specialized in raising Merinos. In the 1980s, my maternal grandparents herded

a flock of approximately 350 Merino ewes. They had to adjust their practices to suit the breed, which is not accustomed to harsh winters. They did so by adjusting the mating period to ensure ewes gave birth in the last two weeks of March. Merino requires assistance in delivery, and lambs are born with a thin coat, requiring immediate sheltering. This called for iron stoves inside the shelters, where herders would work through the night, delivering lambs. Due to these challenges, the herders at Choduraa have since replaced their Merino with the hardier Tyvan breed.

26. The herders put much effort into the preparation of dry dung for winter use. The dung was neatly gathered in several locations at the winter campsite. Before leaving the winter place, the dung was stored under the roofed parts of corrals. In summer the herders would visit their winter camp to pull weeds and to break up the dung piles to aerate them for consistent drying.

27. I am reluctant to use the English "to break a horse." In Tyvan the practice is referred to as *maldy ööredir i*, literally, to "teach a horse." Tyvan herders appreciate spirited horses. They teach their horses to carry a rider, and they do not aim to break the animal's will or make it docile.

28. For details, see the essay-film "Nonhuman Attunements: Tethering Foals and Praising Masters of the Altai Mountains in Mongolia" by the author and Robert O. Beahrs, 2019.

29. Herders who spend summers in the taiga boast about the clean water and green grass of Tangdy-Uula. In their view, the mountainous taiga is best suited to grazing horses in summer. Shara-Nuur herders argue their semi-desert summer grounds, with their nutritious grasses and healing mineral waters, are best suited to training racehorses.

REFERENCES

Anderson, David George, Jan Peter Loovers, Sara Asu Schroer, and Robert P. Wishart. 2017. "Architectures of Domestication: On Emplacing Human-animal Relations in the North." *Journal of the Royal Anthropological Institute* 23, no. 2: 398–416. https://doi.org/10.1111/1467-9655.12613_1.

Anderson, David George. 2000. *Identity and Ecology in Arctic Siberia: The Number One Reindeer Brigade*. Oxford: Oxford University Press.

Arakchaa, Lilia K. 2015. *Etnoekologia Tuvy [Ethnoecology of Tyva]*. Kyzyl: TuvIKOPR SO RAN.

Bavuu-Surun, Mira V. 2018. "Istoria Formirovania Dialektov i Govorov Tuvinskogo Iazyka" [The History of the Formation of Dialects and Subdialects of the Tyvan Language]. PhD diss., Institute of Philology, Siberian Branch of the Russian Academy of Sciences.

Beahrs, Robert Oliver, and Victoria Soyan Peemot. 2019. *Nonhuman Attunements: Tethering Foals and Praising Masters of the Altai Mountains in Mongolia*, A Video Essay.

Beahrs, Robert Oliver. 2017. "Nomads in the Global Soundscape: Negotiating Aesthetics in Post-Soviet Tuva's Traditional Music Productions." *Novye Issledovania Tuvy*, no. 2: 59–110. https://doi.org/10.25178/nit.2017.2.3.

———. 2019. "Tuning a Throat Song in Inner Asia: On the Nature of Vocal Gifts with People's Xöömeizhi of the Tyva Republic Valeriy Mongush (b. 1953)." In The Oxford Handbook of Voice Studies, edited by Nina Sun Eidsheim and Katherine L. Meizel. Oxford University Press.

Donahoe, Brian. 2012. "'Trust' or 'Domination'?: Divergent Perceptions of Property in Animals Among the Tozhu and the Tofa of South Siberia." In *Who Owns the Stock? Collective And Multiple Forms of Property in Animals,* edited by Anatoly M. Khazanov and Günther Schlee, 99–120. New York: Berghahn Books.

Fernández-Giménez, Maria E. 2002. "Spatial and Social Boundaries and the Paradox of Pastoral Land Tenure: A Case Study from Postsocialist Mongolia." *Human Ecology* 30, no. 1 (March): 49–78. https://doi.org/10.1023/A:1014562913014.

Humphrey, Caroline. 1974. "Inside a Mongolian Tent." New Society 31 (October 1974): 273–75.

———. 1983. *Karl Marx Collective: Economy, Society and Religion in a Siberian Collective Farm*, Cambridge: Cambridge University Press.

———. 1995. "Chiefly and Shamanist Landscapes in Mongolia." In *The Anthropology of Landscape*, edited by Eric Hirsch and Michael O'Hanlon, 135–162. London: Routledge.

Kuular, Shomaadyr D. 2002. *Baglaash [Horse Tethering Post]*. Kyzyl: Tuvan Publishing House.

O'Brien, Matthew J., and Todd A. Surovell. 2017. "Spatial Expression of Kinship Among the Dukha Reindeer Herders of Northern Mongolia." *Arctic Anthropology* 54, no. 1: 110–119. https://doi.org/10.3368/aa.54.1.110.

Pedersen, Morten Axel. 2016. "Moving to Remain the Same: An Anthropological Theory of Nomadism." In *Comparative Metaphysics Ontology After Anthropology,* edited by Pierre Charbonnier, Gildas Salmon, and Peter Skafish, 219–243. New York: Rowman & Littlefield International.

Potapov, Leonid P. 1969. *Ocherki Narodnogo Byta Tuvintsev [Sketches of Tuvan Traditional Life]*. Moscow: Nauka.

Stépanoff, Charles. 2012. "Human–animal 'Joint Commitment' in a Reindeer Herding System." *Hau: Journal of Ethnographic Theory* 2, no. 2 (Fall): 287–312. https://doi.org/10.14318/hau2.2.015.

———. 2017. "The Rise of Reindeer Pastoralism in Northern Eurasia: Human and Animal Motivations Entangled." *Journal of the Royal Anthropological Institute* 23, no. 2 (May): 376–396. https://doi.org/10.1111/1467-9655.12612_1.

Tatarintsev, Boris I. 2004. *Etimologicheskii Slovar' Tuvinskogo Iazyka [Etymological Dictionary of the Tyvan Language]*. Vol. 3. Novosibirsk: Nauka.

———. 2015 [2000]. *Etimologicheskii Slovar' Tuvinskogo Iazyka [Etymological Dictionary of the Tyvan Language]*. Vol. 1. Abakan: Cooperative Journalist.

Vainshtein, Sevyan I. 1980. *Nomads of South Siberia: The Pastoral Economies of Tuva.* Translated by Michael Colenso. Edited by Caroline Humphrey. Cambridge: Cambridge University Press, 1980.

Part II

TOFAS, TOFALARIA, IRKUTSK OBLAST'

Chapter 4

People, Reindeer, and All the Others

The Shared Taiga as Common Tofa Household

Konstantin B. Klokov

Translated from Russian by Alex Oehler

The economic history of the Siberian taiga is a story not only of people, but also of the diverse species associated with them.[1] In the taiga, where conditions for farming are unfavorable, both Indigenous peoples and Russian colonists have made their living by hunting wild animals and breeding domesticates. Changes in the political and social structure, economic reforms, and the spread of technical innovations have changed the fate of people, as well as of wild and domestic animal populations. Fluctuations in the abundance of animals, in turn, have determined the structure of the economy and the level of well-being of all those who populate the taiga.

Even today, the life of small taiga settlements depends on the wildlife resources of primarily fur-bearing animals and ungulates. Most of the adult men in such villages are hunters and descendants of formerly nomadic Siberian peoples and Russian settlers. Frequent contact and joint occupation involving some of the same skill sets has led to the formation of a common hunting practice shared among Russian and Indigenous Siberians. Nonetheless, differences in how people relate to domestic animals have remained, primarily because Russian settlers never fully adopted reindeer herding. The main allies of people in the Siberian taiga have been, and still remain to this day, the domestic reindeer and the dog. Horses have also featured importantly.

In Siberia, many local communities emerged where people, dogs, domesticated deer, and horses resided in close proximity to sable, red deer, musk deer, elk, wolves, and bears. Together, they may be said to have formed a large common taiga-based household. This chapter is devoted to the description of contemporary practices that continue to enable such a way of life. It also seeks to describe the particular types of relations that have developed between people and wild and domestic animals in the village of Alygdzher, the largest of three settlements in Tofalaria, a region of Irkutsk Oblast'.

Materials for this chapter were collected through interviews with local residents, mainly Tofas, during two trips to Alygdzher and its surroundings in May 2013 and September 2014.[2] In addition, the chapter draws on an extensive review of ethnographic literature to help illuminate Tofa relations with particular wild and domestic species. Among key ethnographic references are those of B. E. Petri (1927a, 1927b, 1928, etc.), L. V. Mel'nikova (1994), and I. V. Rassadin (2000). We will now turn to the people, wildlife, and domestic animals who reside in the Tofa taiga.

TOFAS

Tofas belong to a relatively young ethnos. They are closely related to Tozhus (or Tyvans-Todzhins of eastern Tyva), having come to identify distinctly as Tofas only in the mid-nineteenth century (Vainshtein 1980, 86). Tofalaria is a remote mountainous region, covering an area of 21,426 square kilometers, located in the Central Saian Mountain Range with elevations of 1,600 to 2,924 meters above sea level. The three main villages in Tofalaria are Alygdzher, Verkhniaia Gutara, and Nerha, all of which are located more than 300 kilometers from the nearest town, Nizhneudinsk. In winter the villages are connected by temporary roads leading over frozen rivers, while for the rest of the year access is limited to aircraft. A total of 1,200 people reside in Tofalaria, of which about half identified as Tofas, and the other half as Russians. According to V. P. Krivonogov (2016), the number of Tofas within their ethnic territory has remained relatively stable in recent decades.[3] In 2015, 535 people lived in Alygdzher, of whom 237 identified as Tofas. An intensive process of métissage and acculturation has been ongoing: 88.4 percent of Tofas are descendants of mixed marriages, and the ancestral Tofa language has been disappearing at a quick rate. In Alygdzher, 78 percent of Tofas no longer speak Tofa, and only 3.3 percent remain fluent. Children under the age of ten no longer understand the language (Krivonogov 2016).

However, a number of Tofa terms relating to reindeer herding remain preserved, and they continue to be used by all the residents of Alygdzher. Among these terms, during my visits, were the following age and sex designations for reindeer: *anai*—calf [*anaichik*—Rus. diminutive]; *ingen*—cow; *chara*—a castrated working deer; *munduchak*—a two-year-old female, which has not bred); *khony*—a two-year-old male; *khudy*—a three or four-year-old male. People also used Tofa terms for traditional objects that lacked a Russian analogue:

solbak—a tethering pole for deer; *taiak*—a staff used for riding deer; *chatyr*—a reindeer corral; *sut*—reindeer milk. A number of Tofa words related to hunting also remained in use, for example: *askyr*—the male musk deer.

As late as the nineteenth century, Tofas nomadically moved through the mountainous taiga of the Central Saians, hunting game and breeding domestic reindeer for transportation. Tofa reindeer herding falls under the Saian Type, which is characteristic only of them and of eastern Tyvans (Vasilevich and Levin 1951). The Tofa (or Saian) breed of reindeer is known as the largest and hardiest of reindeer breeds (Pomishin 1981, 113). In the nineteenth century, a Tofa hunter would rely on three to four reindeer for transportation during a hunt that would last several consecutive months. In those days, Tofa families perpetually migrated, moving camp fifteen or more times per year. In winter, they lived in river valleys to which they descended in autumn, and in the spring they moved to higher ground again. Summers they spent on flat mountain tops and plateaus where the wind blew away midges (Rassadin 2000). In addition to reindeer, the Tofas also held a smaller number of horses, which, like the deer, grazed freely in the taiga (Vasil'ev 1910). Due to their free grazing, a significant number of both deer and horses were lost each year. The Tofas would restore their livestock by purchasing animals from neighboring Soiot and Buriat communities (Rassadin 2000, 34).

With the advent of Soviet power, the economic life of the Tofas changed significantly. The formerly nomadic population was transitioned to a sedentary lifestyle, settled into villages and made to join collective farms [Rus. *kolkhoz* (singular)]. In 1929, the Tofas collectively held 1,213 reindeer, including 426 trained riding deer, and 140 horses (Pervukhin 1930, 84–85). In the following year, all these animals had become the property of three newly formed collective farms. The collective farms, which continued to practice reindeer breeding as a means of providing transportation to their employed hunters, ventured into a new direction: the production of reindeer meat. To be able to export reindeer meat, the number of deer held by the kolkhoz was nearly doubled. The reindeer were guarded by herders who continued to live in the taiga, occupying conical lodges with their families, thus maintaining many aspects of their traditional way of life (Rassadin 2000).

By the late 1960s and early 1970s, the collective farms were merged into a single enterprise—the Tofa *koopzverpromkhoz*, which continued to engage in hunting and reindeer herding. In 1991, after Russia's transition to a market economy, the *koopzverpromkhoz* was reorganized into a closed joint-stock company, which declared bankruptcy in 1999. Its reindeer herd, which in 1993 counted more than 2,000 head, fell to 969 head by 1997, after which its seasonal accounting ceased. During these years, the Tofas endured a difficult period in which it became impossible to earn wages, and all outside supplies of food and goods had been disrupted (Rassadin 2000, 195–213).

Over the past fifteen years, when the Russian economy recovered from the shock caused by the reforms of the 1990s and gradually shifted to market mechanisms of self-regulation, the lives of Alygdzher's residents changed for the better. Ever fewer young people moved from Alygdzher to the city, attracted by the significant profit that could be made from hunting sable and various ungulates in the taiga.

At the time of my visits, much of the food available in Alygdzher consisted of potatoes and other vegetables grown by its residents, combined with game meat, which, in violation of established laws, was hunted almost all year-round. Game was transported from the taiga to the village on horseback, as well as by reindeer. Based on a survey I conducted in 2014, there were about 100 subsistence hunters in Alygdzher. Of them, forty-six relied on reindeer for transportation. These hunters belonged to twenty different Tofa families, some with as many as five hunters, while the majority had one or two hunters among them.

In addition to reindeer, each family was in possession of three to four horses, which, along with the deer, were used to transport cargo to their hunting grounds and to bring back to the village wild meat. Both species were also used to transport hay and firewood from the taiga. The horses grazed freely in herds lacking all human supervision. Their glades were predominantly located in the Kara-Buren valley, about 10 to 15 kilometers from the village.

The purpose of reindeer husbandry was now predominantly to support the villagers' subsistence hunting activities. As I. V. Rassadin (2000, 157) notes, traditional Tofa hunting had largely been lost by the end of the twentieth century. Due to a decrease in the number of domestic reindeer, hunters had begun to ride into the taiga on horseback. Meanwhile, a good number of hunters were being shuttled to distant territories by state-subsidized helicopters. Having abandoned their traditional wooden trap designs, many Tofas began to hunt in the Russian style, relying on rifles and factory-made traps. In place of the hunting traditions once kept by Tofas, the material culture and hunting expertise of Russian settlers gradually took hold.

Following the collapse of the Soviet Union, the entire territory in which the residents of Alygdzher hunted for sable and various ungulates was divided into hereditary hunting grounds. These were locally referred to as taigas [Rus. *taigi* (plural)]. Each family had their own plot on which one or several cabins were located at a distance of about 10 kilometers, one from another. Although both reindeer and horses were used to move between these camps, the latter were far more common. This had to do with the fact that horses could be loaded with more cargo. Also, horses were constantly grazing near the village, making it easier to catch them. Finally, almost all the residents knew how to handle horses, but not everyone was skilled in riding deer. Yet, there were significant advantages to reindeer. Unlike horses, reindeer could traverse through very difficult territory. A number of taigas were located in areas that could be reached only by reindeer or helicopter. Furthermore, horses followed set paths in the mountainous taiga, while a reindeer could be ridden in a straight line through treacherous terrain, chasing sables or checking traps. The off-trail worthiness of reindeer also came in handy when meat from hunted ungulates needed to be carried out of remote spots. Finally, reindeer did not require a supply of oats to be carried along in the taiga in winter. They could feed on lichen year-round.

Reindeer not only carried the hunter, but they also assisted him in the hunt. Possessing a better hearing sense than humans, a reindeer could make out the distant barking of a dog that had discovered a sable, even under windy conditions. Following its fine sense of hearing, the deer would swiftly carry the rider to the site of the dog, taking the most direct route possible (Pomishin, 1971, 129).

Many hunters also used horses and deer in combination. They would move into the taiga in two steps: after the majority of their cargo had been delivered to their winter camp on horseback, the hunter would return to the village, where he would leave his horses, only to travel on with his reindeer. In the middle of winter, when thick ice had formed over the rivers, many hunting grounds could also be reached by snowmobile. However, it was impossible to transport the bulk of one's winter equipment on a snowmobile before the freeze-up of mountain rivers. For this reason, horses, deer, and helicopters remained a necessity. Here are two brief accounts from hunters regarding their winter activities:

> We usually leave to go hunting between October 10 and 15. We travel in teams of two or three people, since on the way we must spend the nights in cabins, but these are small. It takes us about five days to reach our taiga, which amounts to approximately one hundred and twenty kilometers. If three people are traveling, the caravan consists of nine reindeer and six to nine horses. People will ride on the horses, so as not to tire the reindeer before the hunting begins. Reindeer carry only a light load.
>
> I hunt until the end of January without returning to the village. It is difficult to go to the village in winter. To do so, one has to cross a mountain pass with a lot of snow and ice.

Another hunter said, his taiga was located on Shibit River—following a straight line of about 90 kilometers to the southeast of Alygdzher. Were he to follow existing trails, the distance would increase to about 120 kilometers. In autumn, riding on horseback or reindeer, the trek would take some three to four days. In winter, when the snow was deep, it took around five to six days. Once arrived in the Shibit basin, there were seven hunting cabins (for an example of a typical cabin, see figure 4.1). Eight people hunted in the area;

Figure 4.1 Tofa Hunter's Cabin, Tofalaria. *Source*: 2014, K. B. Klokov.

all of them related to each other. As a rule, two hunters would share a cabin, although in some cases a single hunter might occupy a cabin. About once a week, hunters visited each other.

Each hunter would leave for the winter hunt with three reindeer. These he used on a rotational basis: riding one deer at a time, he would allow the other two to rest. In addition to the deer, each hunter also brought along two horses to carry loads. At the end of the hunting season, these horses would help pack out equipment and harvest to the village. The horses did not work over the winter period. Instead, they would graze freely. The two horses of one of the hunters from Shibit River were called 'Blue' [Rus. *Goluboi*] and 'Torn' [Rus. *Rvanyi*]. The latter had received its name from a bear attack it had suffered as a foal, in which its rear end had been severely torn up. Both of them were about ten years old and castrated. Like many other hunters, this man believed horses were smarter than deer. In the past, he had also used a Buran (a Russian snowmobile make) until it broke down. The snowmobile was used only for limited transportation of firewood and wild ungulate meat, as the rough conditions of the mountainous taiga limited its use significantly.

Around New Year, the hunters would travel to the village for about ten days, before returning to their taigas. Accounting for the time spent traveling between the village and the hunting camp, the hunting break lasted for about twenty days. Hunters who did not have their own reindeer were delivered to their taigas by helicopter. On their return from the taiga at the end of the season, they had to walk on foot. The expense of the helicopter was covered by one of the regional social programs. Helicopter fees were too expensive for most hunters to afford on their own.

I will now turn, in greater detail, to how the residents of Alygdzher negotiated their relations with some of the more important wild and domestic species.

PEOPLE AND RED DEER

Manchurian wapiti (*Cervus canadensis xanthopygus*), or elk, was the most sought-after animal by the Tofas. It was the only species that they referred to simply as the "animal" [Rus. *zver'*]. Tofa folklore mentions a shaman riding a wapiti (Katanov 1891). Wapiti, it was said, "feeds" the whole village with its meat. During the winter, each hunter would shoot at least one "animal" for his own subsistence. But hunting wapiti was not merely a matter of feeding oneself, it was also part and parcel of Tofa cultural identity. The "animal" was hunted exclusively with a rifle. Snares and other types of traps were never used on it. The traditional autumn hunt took place during the rut, and it was carried out with the aid of a deer horn, locally referred to as *dutka* [Rus. for pipe]. In fact, this was not just a hunt, but a dialogue between the hunter and his prey. I. V. Rassadin (2000, 158) gives the following description of this traditional hunt:

> In autumn, sometime around mid-September, when the leaves turn yellow, the rut sets in for red deer. At this time, Manchurian bulls begin to bugle, challenging rivals over attempts at their harem. Hunters will use the traditional Tofa *murgu* horn to lure bugling bulls. Usually two hunters are involved; the one plays the horn, the other shoots.

A detailed description of the murgu horn is given by L. V. Mel'nikova (1994, 46–47). In the past, these horns had been made of cedar wood. More recently they were fashioned from a strip of film, used at the hospital for radiographs. The film was rolled up into a tube, after which an air-reed made from birch bark was inserted into one end to lure musk deer (see below). Lacking a horn, hunters would also bugle through the barrel of their rifle, or simply imitate the bull's call with their own voices. If at the beginning of the rut wapiti were cautious, then by the end of it—after the bulls had already impregnated several females—they would lose all sense of caution and run toward "any piping."

Moving through the taiga in autumn, hunters would, every now and then, sound their horns to locate and mark in advance the position of bulls in the landscape. This was not yet part of the hunt, but merely a kind of roll call. When hunters crossed paths in the taiga, the usual topic of their conversation was who, where, and when had responded to these calls. The hunt itself began only later. By that time, hunters already knew where the "animal" was located, and how it was responding to the murgu horn.

The hunt began with the hunter sounding his horn until he heard a response from a red deer. If the answer came from the windward side, the hunter would immediately approach the animal against the wind. If the wind came with the hunter, he would walk around the animal, to approach it against the wind. Approaching the bull, hunters purposefully made noise, breaking branches while remaining unconcealed. Sometimes an agitated bull would approach the hunter all by itself. In this case, the hunter would make more noise, breaking branches to further agitate the animal, causing it to come closer.

The rack of antlers of a large bull were set aside as a trophy. Its age and size were determined based on the number of branches. Hunters would speak of "a bull on eight," or a "a bull on twelve," referring to the number of antler points. "On eight" meant four branches on each antler; "on twelve" would be six on each side, and so on. Antlers and dry horn were sold to dealers in the city. The rack of a red deer sold for 600 rubles per kilogram. Dry horn sold for 60 rubles per kilogram. Meat could not be sold, as there was no demand for it in the village, and police prohibited it from being transported to the city by helicopter, since it had been hunted out of season and without a license.

PEOPLE AND MUSK DEER

In contrast to red deer, musk deer (*Moschus moschiferus*) had more recently become the single most important source of cash flow. Hunters were regularly selling its musk pods [Rus. *struiki*], a sale that usually took place illegally. The importance of hunting musk deer had increased following the drop in prices obtained for sable pelts, which had previously made up the majority of a hunter's income. The hunt on musk deer took place year-round, and animals were most commonly caught with steel cable snares. For this purpose trees were felled and arranged to create barriers [Rus. *zaseki*] containing two, three, or four openings into which the snares were installed. Tree barriers measured 100 or more meters in length, and they were usually placed between two rocky outcrops.

Naturally, with this method, the meat of prey often deteriorated before a hunter could get to it, but this mattered little since the main goal was to obtain the musk pods. I will provide a brief excerpt from a typical account offered by a musk deer hunter; an account that would be difficult to follow without the context I just gave: "A hunter approaches us, and he stinks of carrion. I ask him, 'Were there any struiki'? He answers, 'I took two struiki.'" It is clear from the conversation that the hunter went to check on the snares he had set, in which he had found two decaying musk deer.

Because this method of hunting was illegal, hunters avoided talking about it. Its scale could only be ascertained based on occasional observations. One hunter, for instance, was loading 300 meters of steel cable into a helicopter on his way into the taiga. Another hunter, together with his colleague, said they had harvested eighty musk pods over the course of one year. During the winter, many hunters collected up to 1 kilogram worth of musk glands. According to some hunters, a male musk deer (locally known as an *askyr*) carried between 15 and 50 grams of musk pod, averaging around 20 grams. To obtain a kilogram, one needed to trap about forty males. Prices for the pods were quite high, which made hunting for musk deer no less profitable than the hunt on sable had been. However, hunters became dependent on intermediary dealers, who ilegally resold the pods to China and other Asian countries. In order to legally sell musk pods, a hunter had to obtain a license, but licenses were limited in an effort to protect the endangered species.

According to several hunters, the musk deer population had decreased in recent years, since the taiga had become covered with snared tree barriers, which could be spotted even from a helicopter. In Alygdzher alone, about 100 licenses for musk deer were being given out yearly. The hunters themselves estimated their prey at about 400 musk deer. Approximately the same number of musk deer had been harvested here in the 1920s (Petri, 1928, 24–26). However, in the past musk deer had been harvested primarily for their meat, and their musk pods constituted merely 2.2 percent of the total annual hunting income (Petri, 1928, 24). Now mainly male musk deer were being harvested to obtain the prized struiki.

Besides hunting musk deer with snares, which was somewhat of a commercial activity—not to mention its poaching quality—hunters also obtained musk deer following a more traditional method that made use of the *pekulka* whistle.

In the spring, when the snow melted and does were having their fawns, as well as in summer, when their young ones were growing up and lying in hiding more often, hunters would lure musk deer with the *pekulka* whistle which imitates the cry of a hungry kid (Rassadin 2000, 158). The whistle was made of birch bark, and it could quickly be fashioned in one or two minutes. One sample was made right in front of me. Hunting in this way was referred to as *pikat'* [Rus. verb form of the noun pikulka]. It was best to pikat' musk deer in spring—from late May to early June—when many of the males responded to the sound. It is unclear why not only females are drawn out by the sound. However, according to hunters from Alygdzher, musk deer bulls were also suckling milk from the females at this time of the year. The female, according to them, would run immediately "under the feet" of the hunter, while males were more cautious, usually stopping off to the side and listening. Consequently the males often stayed out of reach of the hunter. Hunters would try to shoot only males, but it was difficult to distinguish a male from a female on the run—it took some time to note whether the animal had fangs or not.

In the past, other now nearly forgotten methods were used to hunt musk deer. Hunters had used their dogs to drive musk deer onto steep rocky outcrops [Rus. *otstoi*], and when snow covered the ground, they would track the animal to its daytime bedding.

PEOPLE AND WILD REINDEER

Wild reindeer were not generally hunted, as they roamed too far from Alygdzher. However, it did happen that a wild bull reindeer would come into a domestic herd during the rut. These bulls had sharp horns and could cause damage to the domestic deer, whose tines had been clipped. Herders usually shot such bulls for meat. However, according to some hunters, it is better to first allow the wild animal to spend some time with the domestic cows to diversify the herd's gene pool. The offspring of wild reindeer in a herd of domestic animals was easily spotted. Their coat color was generally lighter, and they were overall more enduring, of a stronger build, and less docile.

Opinions on whether it was a good idea to mate female cows with wild bulls differed. Some believed refreshing the herd's genetic pool was a good thing. Others objected, pointing out that the character of the offspring of wild animals resulted in equally wild demeanor. "We call them *dikashary*. [In the past] they have often grown up to become semi-wild, and if such an animal was not trained, it would be slaughtered. Often it happened that these animals left the herd on their own." The calves of such unions were kept on a tether from their early days on, in the attempt to domesticate them. However, this was not always successful. "You can't drag the offspring of a wild reindeer anywhere. They are difficult to teach." "A cow descended from a wild bull will run with its calf following behind. When tethered, they will spin wildly, sometimes to the point of becoming entangled by the rope, and choking themselves to death. They will also try to jump the fence of a corral, where they can get stuck, getting hanged."

PEOPLE AND WOLVES

Tofas had a particular attitude toward wolves, as is the case with many Indigenous peoples of Siberia. In the past, its name had been a taboo, and people had referred to it only as *kuduruktukh*—"the one with a tail" (Rassadin 2000, 50). As B. E. Petri (1928, 22–23) noted, Tofas traditionally did not hunt wolves, neither did they set traps for wolves or use poison. A deer killed by wolves was not eaten, as it was considered the prey of a wolf (Petri 1927a, 21). If wolves came to attack domestic reindeer, Tofas quickly withdrew from the scene, attempting to move as far as possible to prevent the pack from following the household. If several families of reindeer herders had been camping together, they would each depart in a different direction (Petri 1927a, 19). While attitudes toward wolves had changed in more recent years, some of the older sentiments toward wolves remained.

Over the past few decades, with the number of wolves increasing throughout Siberia, and with government measures to combat predators remaining insignificant, wolf attacks had become a serious threat to taiga reindeer herding in general. However, at Alygdzher, hunters and herders did not consider wolves a serious threat, despite the free grazing of their reindeer and horses. I was told about a lone wolf who resided near the autumn reindeer camp, as well as a pack of five wolves who lived nearby. Whether these wolves made attempts on domestic deer was not clear, since the reindeer were grazing freely and lacked constant supervision. However, everyone in the village was aware that in 2014 the wolves had killed a horse, which had been grazing alone for several months in the taiga. It was characteristic not to blame the wolf for this loss, but the owner, who had left the horse unattended for such a long time. No other incidents had been reported. Two wolves had recently been killed near the village. They had gotten caught in snares set for musk deer. Prior to this, no one had hunted wolves for a long time.

Judging by these events, the wolf population of Tofalaria may have been on the rise, posing an increased threat to reindeer and horses. In the meantime, however, the residents of Alygdzher were not alarmed: "They are not around the village, they haven't come out for a long time," and "there are many wolves in the mountains, but they keep busy with the red deer and don't bother us." Perhaps, this respect for wolves, which is so characteristic of many peoples of Siberia, was the result of a combination of the desire to be at peace with the species and the fear that they would take revenge for anything done to them.

Hunters remembered how, in the Soviet years, considerable attention had been given to the fight against wolves. Special traps had been set for wolves and lures laced with strychnine had strategically been placed for them. Poison was then the main way to fight wolves, not only in Tofalaria, but also in other areas. At time of visit, the sale of wolf poison was prohibited, and all the old stock had been used up. Hunters also felt uneasy about using wolf poison, as they thought it dangerous to skin a contaminated wolf, lest the hunter be poisoned himself. Hence, hunters left the skins of poisoned wolves in the taiga, which meant they could not turn in the skin for its prize money. In the past years, turning in wolf skins had been lucrative, but now the premium per wolf was only 3000 rubles. This was not enough to interest a hunter in a wolf as prey.

PEOPLE AND DOGS

There were a lot of dogs in Alygdzher, and these were exclusively hunting dogs. There were no reindeer guiding dogs or sled dogs. Only one hunter harnessed two of his dogs to draw a children's sled for cargo. There were no special guard dogs in the village either, but hunting dogs often performed a guarding function. According to local residents, "Half of the hunting dogs in the village are biting dogs." A good hunting dog was well cared for, even if it bit people or stole food. People would rather tie it down than kill it. By contrast, dogs in Alygdzher never seemed to pester horses or reindeer.

Ungulate and sable hunters kept three to four dogs on average. Training a hunting dog took little effort, as the dog would learn from the first day it was taken on the hunt. "There are no special techniques for training hunting dogs, you need to take the dog into the taiga and walk with it," the hunters said. A dog either picks up on the skills of another more experienced dog, or it guesses what must be done by observing the behavior of the hunter. A hunter only interferes with a dog's wrong actions. If an untrained dog does poorly chasing a sable, for instance, by repeatedly losing sight of it or by giving up on the sable, the hunter would call the dog in and re-direct it, ever urging it on to try to maintain its interest in pursuing prey. Sometimes, in order to train the dog to look for sable on the trail, the hunter would put the dog on a leash, following it along a fresh trail until the dog located the sable.

Here is a hunter's brief account of hunting with his dog:

> I just show the dog a fresh track and start urging it on, or I let it work alongside a more experienced dog who will teach the novice, thereby ensuring the sable will not deceive the learner. I usually have two dogs working together on a sable. The sable hides from the dogs in a *sharlop* [local term for scattered rocks or scree on a slope] or in *poddor* [voids formed by a dry creek], or inside a hole leading into the hollow of a thick cedar tree. We smoke it out from such cavities, although this can be difficult or even useless in a large sharlop. A sable hunting net does not apply here either. Once the sable pops out, the dog is supposed to chase it, not allowing it to return to the scree. Often a sable will gather

the dogs into a single location by descending from a tree, only to leap over the dogs and run.

In the 1920s, when there were a lot of squirrels and fewer sable, dog training looked somewhat different. At first, dogs were trained to track squirrels, as it is much easier to work with squirrels than with sable. Only after mastering the squirrel were dogs allowed to hunt sable. Hunters were of the conviction that a good squirrel dog would progress to hunting sable all by itself, whereas a poor hunting dog would stick with squirrels (Petri 1928, 32).

People's attitude toward dogs in everyday life had also changed. In the past, when Tofas lived nomadic lives, a good hunting dog was valued as highly as two horses. A dog was almost a member of the family. In winter it would sit by the fire, alongside people, and even near sacred objects inside the conical lodge (Petri 1928, 33). Dogs also actively participated in the upbringing of children. Popov (1878, 3) describes Tofa dogs in the family not only as faithful and reliable friends, but also as caring nurses. Holding on to the dog, a Tofa child would learn to walk. "Often [a child] is saved from falls, bruises, and other misfortunes common to an early and neglected childhood." Tofas ended the lives of aging dogs with special honors, shooting it in "a good place somewhere on a mountain," after feeding it and wishing it farewell with the words: "go on now to your place" (Petri 1928, 33). At present dogs were no longer allowed into people's homes, and few could be seen running around the village. Instead, they were kept inside fenced yards, and many of them were aggressive toward people. Retired working dogs were usually shot, although many hunters would delegate the task to someone else.

TOFAS AND DOMESTIC REINDEER

At the end of the twentieth century, Tofa reindeer herders had retained several of their traditional reindeer herding practices. They were still milking the cows, using their milk for food, and they rode them in winter. To count and vaccinate them, the deer were driven into corrals (chatir) made from poles and thin logs, just as had been the practice in nomadic life. The design of these corrals had not been altered. Reindeer intended for hire during the winter hunt, were kept in a separate herd. Only the strongest animals were selected and castrated for this purpose. The weak, the sick, and the old were fattened in another herd for meat. In Autumn, on their way out to the winter hunt, each hunter would rent several reindeer and horses from the common herd for the duration of his winter activities. Such was the reindeer herding of Tofas during the Soviet period, and it remained so until the end of the twentieth century.[4]

Some of these features had stayed intact, although the number of deer had decreased threefold over the past fifteen years. During my visit, there was only one deer herd in Alygdzher, counting roughly 300[5] adult deer. The herd was looked after by two herders. Over the past ten years, the herd had been grazing east of Alygdzher in the headwaters of the Dugul'ma River in autumn, winter, and spring. During the calving period in spring, another two or four people would come to assist the herders. They were called calvers (Rus. *teliatniki*). In the past, elderly women had worked as calvers most commonly, but at present all the calvers were men from the village in their mid-twenties and thirties.

Calving began in early May and ended at the beginning of June. At the end of April, about a week prior to the onset of calving, the pregnant cows (Tof. ingen) were tethered to a *salbak*. The salbak was a horizontal tethering pole made from the stem of a debranched tree, 10 centimeters in diameter. The ingen was then tied to the salbak with a rope, approximately 1.5 meters in length.

Shortly after the calf [Tof. *anaichik*] was born, it would be tied to the salbak with the help of a bridle that was attached to a wooden *mungui* (see chapter 7), allowing the animal to turn freely without getting tangled. Following the birth of her calf, the mother was released from the salbak to prevent her from injuring her newborn. She would now graze freely during the night, while being tethered to the roots of a tree during the day.[6] The calf was tethered during the day, while its mother grazed freely. Two weeks later, the calf would be allowed to graze freely at night, while its mother was tethered. Thus, mother and calf were taking turns grazing freely. During the night, the ingen would wander as far as 2 kilometers from her anaichik, but she would always return by morning.

In recent years, about seventy calves had been born in the herd each year, but only half of them survived until autumn. According to the residents of Alygdzher, calves mainly perished from predator attacks, primarily from wolves, but also from bears and wolverines. All spring, the calves and cows were tethered on rotation, until, at the end of June, the herd was driven toward the snowfields at higher elevation, which would serve as their *letnik* (Rus.), or summer pasture. Here the entire herd roamed freely. The calvers would depart at this point, leaving the herd to two herders.

By early September, the herd returned to the Dugul'ma Valley, where it remained for the winter until the next calving period. During the cold months, the herders were using a small log cabin, but for the rest of the year they were staying in tents or conical lodges covered with tarpaulin.

In late September, just prior to the rut, the herders would trim the antlers of all the males in the herd. The antlers of reproductive bulls were cut down to the pedicle to prevent them from butting each other. On castrated working deer only some of the tines were clipped to prevent the bridle from getting tangled. The cows' antlers were largely left intact.[7]

Come early October, hunters from the village selected their riding deer before departing on the winter hunt. Two herders were now left with a herd consisting of groups of cows, each with their reproductive bull.

After returning from their winter hunting activities, the hunters would release their transportation deer just outside the village, where they grazed free of supervision, until the gadflies and horseflies became too thick. By the end of June, the deer would migrate, on their own initiative, to the summer pastures in the snowfields at higher elevation. By this time, the herders had already arrived there with the remainder of the herd.

To keep the deer tame, and to prevent them from straying too far during this time of free range, the herders were constantly attracting them to camp, using salt. The salt was poured into wooden troughs or simply onto large boulders. Troughs were made from logs into which deep grooves had been chopped or cut by axe or saw to hold the salt. The herd could consume several kilograms of salt per day.

Hunters and herders brought a lot of salt to camp, as they would give it not only to the deer, but also to their horses, and to lure wild red deer. During the hunt, and at camp, people would urinate in the same location, so deer could eat the salty urine-saturated snow. A total of approximately 200 to 250 kilograms of salt was used per hunter per year. Some hunters also brought with them hay or dry feed mixture as supplement feed for their transport deer when necessary.

Transport Deer

The Tofas used working deer during the hunt from late September to late February. Before departing for the winter season, each hunter would come to the herd to collect his deer, which were tagged by an earmark. Tagging by notching or cropping ears was a novelty that had been adopted only in the 1990s. It communicated that a particular deer was in use by a particular hunter. According to the order established by the village administration, for each deer taken from the public herd, a hunter had to pay 500 rubles per year.

Usually, a hunter would select a young male reindeer—most commonly a two-year-old animal [Tof. *khony*]—tagging it with his signature notch. For the first year, the deer would carry only a light load. Once the animal turned three years old [Tof. *khudy*], the hunter would castrate it, using a knife. The health of a transport deer largely depended on the conduct of the hunter using it. For this reason, hunters were expected to adhere to a set of guidelines ensuring deer were not overworked. Going uphill, a hunter would pace his caravan by stopping and resting his animals approximately every 100 meters, or as soon as they got out of breath. A deer that was driven ceaselessly, would become obstinate, unwilling to move at all, even when beaten with a stick. If ridden or handled improperly, a deer's spine could be injured. For example, if a deer tied behind another suddenly jerked with great force on its reins, the leading deer's spine could take injury, which meant the hunter had to slaughter the deer. Overloading could lead to the formation of bumps [Tof. *khala*] around the joints of the deer's front legs. Excessive weight would cause the joints to become inflamed, which resulted in the formation of callus.

A hunter usually took no more than four to six deer on the hunt. In the evening the deer were hobbled by taking the reign from the bridle and tying it to one of the rear legs. They were then released to graze for the night, wandering off for a distance of 1 or 2 kilometers while staying together. Each morning of the hunting period would begin with the hunter rounding up his hobbled deer, which took up an hour on average. Having gathered his deer, the hunter would have some breakfast before riding off for the day.

Reindeer were also used as pack animals, delivering hunting equipment to the taiga and packing out meat and furs afterward. It follows that Tofa deer were rated on the basis of their carrying capacity and endurance. According to one Tofa, there had been a reindeer-riding hunter who had weighed over 100 kilograms. Including his equipment and saddle, his deer would have been carrying about 120 kilograms.

On average, however, a pack deer arriving at a hunter's taiga was loaded with no more than 40 or 45 kilograms of cargo. A grown chara was able to carry this weight for six to eight hours per day. Usually, hunters would travel on their deer for about four hours, after which they stopped for tea. After the break, they would resume their travels for another four hours. In winter, when the days were shorter, less time was spent moving. Young reindeer required a lighter load. A khudy could be loaded with 20 to 25 kilograms, while a khony was to take no more than 10 to 15 kilograms. Yet there were some hunters who rode even their khudy.

Hunters made their own pack saddles for reindeer. A pack saddle's construction was quite simple. Riding saddles, on the other hand, were quite complicated to construct and many people had forgotten how to make them. Most people were still using old saddles that had been preserved. In some cases the old wooden stirrups had been replaced with steel equivalents from horse saddles. In the past, children's saddles had also been fashioned, but at present no children were being transported by reindeer.

Training Transport Reindeer

There is almost no information in the literature about the training of deer by Tofas. Consequently, I paid particular attention to it during my visits. In fact, there was hardly any training to be observed. Instead, animals were being tamed. Part of this procedure was accustoming the deer to calmly walk on a rein with a load or a rider on its back, and to refrain from sudden jerking, from trying to run away, and from dumping its load (see Figure 4.2). The deer were also being accustomed to the presence of hunting dogs. The best way to achieve this taming was for a person to physically engage the animal, repeatedly tethering it or walking it by its bridle, and bringing it into regular contact with other reindeer by tying several deer together and having them move in this way.

Here is a description of how deer were tamed during my visit (Fieldnotes, 2014):

> First they teach them to carry a pack saddle. After saddling the deer, they will walk it on a reign. By the second time the deer can already be linked to the leading deer [in a caravan]. The trainee's reign is attached to the saddle of the leading deer, while another experienced deer may be tied behind the learner. Training a deer for riding is more involved. It is accomplished more easily by two people. One person rides ahead, drawing the learner behind on a rope. The second person sits on the learning deer, urging it on. After some time, the person riding in front passes the reign on to the rider sitting on the learning animal, and the latter now begins to ride the deer independently. After two or three days, the deer gets used to its new rider and it can now be ridden on its own.

Figure 4.2 Unidentified Tofa Herder with Several Chary, Tofalaria. *Source*: 2014, K. B. Klokov.

There was another way of training a deer: by using a staff [Tof. *taiak*]. Most Tofas were sceptical of this method, seeing it more as a joke: "it is useless, although I've tried everything." "A taiak can be used when training a deer without a partner. The taiak is held out in front of the animal [with the reign running from the bridle to the far point of the staff], the reign running in between the two antlers [and back to the rider's other hand]. The person pulls on the reign, and the reign pulls the deer forward. You can also take the reign and the stick in the same hand and pull the reindeer forward in this manner."

Reindeer Character

Tofas claimed that reindeer did not understand verbal commands, but that one could effectively call a reindeer by voice, no matter what words one chose. All that mattered was that the animal heard one's voice. To train deer to respond to the human voice, herders would yell whenever they were scattering salt. The deer would get used to this and approach the herders upon hearing their yelling.

Although it was not customary to name every individual deer, some animals did have nicknames. As a rule, nicknames reflected features of appearance or character. For example, one hunter would take the same three deer on his annual hunt, year after year. He had nicknamed the first one *Bol'shoi* (Rus. for "Big"), the second was named *Krivokhvosti* (Rus. for "Crooked-tailed"), and the third went by *Seryi* (Rus. for "gray," as well as the short form for Sergei). Another hunter had four deer, which were named Nimble [Rus. *Shustryi*], Savage [Rus. *Dikar'*], Hook-nosed [Rus. *Gorbonosyi*], and Fat [Rus. *Tolstyi*]. A third hunter had five deer: Yellow-eyed [Rus. *Zheltoglazyi*], Bulldog, and three younger deer, which had not been nicknamed as yet.

Among the many different external characteristics by which herders identified their deer were the shape of their antlers, their overall size, coat color, the shape of their muzzle (curved or straight, etc.), as well as unique features, such as slanted or curved tails. Additionally, herders had recently begun to trim numbers into their fur. This was convenient, as it allowed the animals to be identified from afar. Deer with numbers were thus referred to as "three" [Rus. troika], "seven," or "eleven," and so on.

Reindeer Dairying

As I. V. Rassadin (2000) notes, Tofas did not usually drink the milk of domestic reindeer directly, but brewed tea with it instead. They had used green tea, which had come in the shape of compressed bricks. Sometimes, fresh milk was also used to prepare a *salamat* [Rus. for a flour porridge or thick flour *kissel* (leavened flour porridge), made from roasted rye flour and brewed with boiling water—or in this case with fresh reindeer milk]. This dish was usually cooked for women in labor. When there was a lot of milk, it was soured and made into cottage cheese. Sour cream was further churned into butter. Approaching autumn, before the end of the lactation period, herders began to accumulate milk for their winter use. To preserve the milk, it was poured into cleaned and dried ungulate stomachs, which were then hung up to dry inside the conical lodge. In this way, the milk did not spoil all winter.

At the time of my visit, the tradition of reindeer dairying was nearly lost. Although deer milking occurred only rarely now, Tofas still recalled its practice in detail.

Cows with calves had been milked once or twice a day, usually in the morning and in the evening. If the cow had been grazing freely during the day with its anaichik tethered at camp, then upon return in the evening, the cow was tethered for milking. Before the she was milked, however, her calf was allowed to suckle for a bit. The suckling was limited to the two smaller front teats, with the milking person covering the two larger rear teats by hand. Subsequently, the calf would be tied off to the side, so as not to interfere with the milking which would include the larger rear teats.

During the night, when the calf had free range, it was able to suckle as much as it desired. In the morning, when the calf was tethered again, its mother would be released. If the cow was to be milked in the morning also, the calf would be tethered for one to two hours prior to milking. This prevented the calf from suckling its mother dry prior to the morning milking session. The herders would tether the calf on a particularly short reign, preventing it from reaching up to its mother's teats, even if the mother were to move close to it. On occasion, only the cow was tethered. In this case, the calf was simply shooed away.

Cows were milked from spring until autumn. In autumn the herders would milk more than in the spring, as they were preparing their stash of dried milk for the winter. In autumn, the cows were milked only once in the morning, but they were now producing a greater and more fatty milk yield than in spring.

Each cow had a daily yield of 150 to 300 grams of milk. Milking ten to fifteen cows per day, a herder was able to obtain anywhere from 1 to 3 liters of milk per day. Especially when only small amounts of milk were available, the milk would be consumed immediately. Most often it was drunk with tea, because of its cream-like fattiness. After boiling, it became thick like sour cream, allowing it to be spread on bread or to be eaten with a spoon. People also made soup, portage, and butter. For these purposes, the milk was heated in a pot, without bringing it to a boil. The pot was then tightly sealed, wrapped in a skin and set aside to sour for two days. On the second day, the separated cream was scooped off the top, leaving the sour milk at the bottom. The cream was then churned into butter. If the cream was transported in saddle bags on reindeer, it often did not require churning as it would turn into butter from the slushing movement.

Milk could also be turned into *pyshtak* (Tof.). For this purpose, the milk was boiled to curdle it. It was then pressed for two or three days, much like cottage cheese. No salt was added. What resulted was similar to hard cheese that could be sliced with a knife. Pyshtak lent itself well to milk collected in summer.

Autumn milk was used to amass winter rations. The milk was poured into a jar where it sat for up to a day, after which it was boiled for about five minutes. During the boil it had to be stirred constantly to prevent it from burning. The milk would thicken while boiling. Afterward the milk was poured back into the jar, where it sat for another twenty-four hours, before it was poured into a *kurdiuk* (Tof.). The kurdiuk was a special sack made from the peritoneum of a musk deer or a reindeer. Here the milk would continue to thicken, turning into a kind of cottage cheese, but without drying out completely.[8] On the following day, another portion of milk was poured on top of the previous filling, which was repeated until the sack was full. A kurdiuk held about 3 liters. In winter, milk would freeze, which made it dry out on the edges. In a kurdiuk it kept all winter, and one could eat it with a spoon, much like cottage cheese with its kernels. It could also be added to tea in this form. More recently, milk preserved in this way has been eaten more commonly with onions or garlic, although in the past it was eaten plain. The kurdiuk did not have to be made from an ungulate's peritoneum. It could also be fashioned from harder materials, but in that case the milk would dry out.

Unfortunately, these traditional methods had largely been discontinued by the time of my visit. The herders would now drink reindeer milk only on occasion and only with their tea. They would milk five or six cows, whenever someone had the desire to drink some reindeer milk.

DISCUSSION

It is interesting to compare the information on human-reindeer domestication relationships described above with data for other taiga peoples of Eastern Siberia, first and foremost for Evenks. Given the large number of commonalities, there also exist a number of differences, which can be explained on the basis of their divergent geographical contexts. Tofalaria is a mountainous country in which only riding and pack reindeer can be used, and where draft reindeer are not an option. In all of Tofalaria's recorded history, there has been only one instance in which reindeer were made to transport cargo on sleds. The experiment occurred between 1942 and 1943. The reindeer, which were unaccustomed to the sleds, quickly broke down, and the sleds were never again attached (Pomishin 1971, 129–130).

Tofalar deer are large and resilient, and in terms of their size and work performance they far exceeded Evenk deer. Evenks also differ in the way they trained their reindeer, which again has to do with the nature of the landscape in which they reside. Given their relatively level taiga, Evenk reindeer herders often made use of sleds. They began by pairing a new draught animal with a more experienced one in the lead. The following year, the new deer was put in lead position. Only after the new deer had mastered this skill—at which time the deer was four to five years of age (one year later than in Tofalaria)—was the it taught to carry a saddle. Only a light person could ride on a four-year-old Evenk deer, and riders of a larger build required a five-year-old deer (Davydov 2014, 108–109). In the Evenk districts of Buriatia and Iakutia, not every riding deer could carry a rider weighing 70 or 80 kilograms, whereas in Tofalaria any working deer was capable of it (Pomishin 1981, 125–126). Yet, Evenki and Tofas held in common the same basic principles of working with reindeer: their main focus was on reducing the degree of wildness in animals. In fact, this was a form of taming, which is part of domestication—an ongoing process which had begun in antiquity and continued into the present (Davydov 2014, 116).

Tofas and Evenks recognized a similar degree of individuality in their reindeer. Both knew their deer by sight, treating them as "other than human persons" (Hallowell 2002[1960]). Since reindeer did not respond to their given nicknames, although they responded to human vocalization, both Tofas and Evenks gave nicknames only to some of them. These were usually bridled working animals. Nicknames were not used to solicit the animals, but to communicate information about them to other people. Nicknames often reflected the disposition or appearance of the deer (Davydov 2013, 36–38; 2014, 106).

Unlike Evenks, Tofas never drove their reindeer into expansive sections of the taiga, sectioned off by fences. Instead, Tofa chary (castrated riding deer) spent several months each year on their own in the taiga, lacking any sort of human supervision. This time of separation did not, however, lead to them going feral in any way. By autumn, these chary would allow themselves to be caught, even without the use of a lasso. Herders of Alygdzher reported that there had been cases in which some chary had failed to join the main herd in summer, skipping a whole year of the herding cycle, before returning the following year. It can be said, thus, Tofa deer were more tame and, at the same time, more independent than Evenk deer. The fact that Tofa herders hardly ever relied on the use of lassos speaks of the high degree of their reindeer's tameness.

Much commonality can also be found between the ways Tofas and other Siberian peoples related to wild animals. The peculiar attitude Tofas possessed in regard to wolves, for instance, may be considered the survival of an ancient ritual context they held in common with other, neighboring peoples.

Evenk reindeer herders of the northern Baikal region (Davydov 2013, 25–26), much like Tofas, refused to consume meat of reindeer that had been killed by wolves. They also respected the lives of wolves, and sought to kill wolves only when absolutely necessary. Yet, both had used poison to kill wolves in the past. A more categorical statement was recorded by V. Davydov (2014, 113) of an Evenk from Tianya (South Iakutia), who recalled: "My grandfather told my father, 'never hunt a wolf, don't scatter poison—Evenks have such a law.'"

I met a similar attitude to wolves among Evenk hunters and reindeer herders in Surinda in the Podkamennaia Tunguska Basin of Krasnoiarsk Krai. Here a bear that had attacked domestic reindeer was killed immediately, whereas herders would move the entire herd to another place if wolves had been attacking. The Evenk herders explained, they would not hunt wolves, but sought to make arrangements with Russian hunters to shoot the wolves. In the past, the herders themselves had used poison against wolves, but at present this method was prohibited and it was difficult to obtain poison.

In southern Chukotka, I was told by some elderly reindeer herders that upon finding a wolf's den, they would never kill the young. Instead, they would tie strong cords around the necks of the pups so they would gradually strangle themselves as they grew up. The herders explained this by saying that their mother would take revenge if she found out that people had killed her offspring. Fear of wolf revenge for the killing of another wolf has also been recorded among Evenks (Tugolukov 1962, 76). One of the most memorable descriptions comes from V. N. Toboliakov (1930, 91–92), who, during a trip to the tundra of northern Gydan in the Russian Far North, met a Nenets reindeer herder who was keeping two young wolves on chains near his conical lodge. The man had been feeding the pups and was planning to release them back into the tundra in autumn. The wolves were to tell their pack mates how well the man had treated them. The herder thought that this would prevent the other wolves from attacking his reindeer.

Many more aspects of the human-animal relations described above lend themselves to comparison with other traditions across Siberia. In uniting people, and wild and domestic animals, each multispecies community possesses a number of its own features, reflecting its unique adaptive strategies to the specific geographical environment in which it finds itself embedded. Meanwhile, life in these communities is also partly based on common traditions, which are aimed at securing a balanced and sustainable existence in a shared landscape.

NOTES

1. Materials for this research were collected during expeditions organized with funds from the Arctic Domus project (ERC Advanced Grant 295458). Data

processing and text preparation was carried out with the financial support of the Russian Science Foundation project No. 18-18-00309).

2. These visits were part of the Arctic Domus research project at the University of Aberdeen, UK.

3. Krivonogov presents the following Tofa demographic fluctuation by year: 1985–598; 1995–661; 2005–665, 2015–663.

4. More detailed accounts of these practices are found in the works of L. V. Mel'nikova (1994) and I. V. Rassadin (2000).

5. Most herders and hunters reported between 200 and 300 head.

6. There is some discrepancy between this account of the tethering sequence and that reported by Oehler, based on Tulaev's account, in chapter 7 of this volume. Likely, this has to do with individual preferences of herders.

7. Depending on the herder, detailed accounts of clipping procedures differed slightly (cf. Oehler, chapter 7, this volume).

8. Russian cottage cheese is generally dry, unlike many of its equivalents in other countries.

REFERENCES

Davydov, Vladimir. 2013. "Bor'ba s Khishnikami i Povsvednevnye Praktiki Sovermennykh Olenevodov: Otnoshenia Cheloveka i Zhivotnykh na Severnom Baikale (Po Rezultatam Polevykh Issledovanii 2007–2012) [The Battle with Predators and Daily Practices of Contemporary Reindeer Herders: Human-Animal Relations in North Baikal (Based on Data Gathered During Field Work 2007–2012)]." In *Materialy Polevykh Issledovanii* [Fieldwork Materials] edited by E. G. Fedorova, 23–42. St. Petersburg: Museum of Anthropology and Ethnography.

———. 2014. "Issledovanie Otnoshenii Cheloveka i Olenia v Iuzhnoi Iakutii [An Analysis of the Human-Reindeer Relationship in Southern Iakutia]." In *Materialy Polevykh Issledovanii* [Fieldwork Materials], edited by E. G. Fedorova, 95–117. St. Petersburg: Museum of Anthropology and Ethnography.

Hallowell, Alfred. 2002[1960]. "Ojibwa Ontology, Behavior and World View." In *Readings in Indigenous Religions*, edited by Graham Harvey, 18–49. New York: Bloomsbury.

Katanov, N. F. 1891. "Poezdka k Karagasam v 1890 Godu [Journey to the Karagas in 1890]." *Zapiski Imperatorskogo Russkogo Geograficheskogo Obshchestva Po Otdeleniu Etnografii* [Papers of the Imperial Russian Geographical Society from the Department of Ethnography] 17 (2): 133–230.

Krivonogov, Viktor. 2016. "Sovremennye Tofalary (Opyt Interval'Nykh Issledovanii) [Contemporary Tofalars (Findings of a Longitudinal Study)]." *Novye Issledovania Tuvy* [New Research of Tuva] 3: 122–145. http://nit.tuva.asia/ nit/article/view/469.

Mel'nikova, L.V. 1994. *Tofy. Istoriko-Etnograficheskii Ocherk* [Tofas. An Ethnohistorical Study]. Irkutsk: Vostochnosibirskoe Knizhnoe Izdatel'stvo [East Siberian Book Publisher].

Pervukhin, I. 1930. "Karagasy [Karagasses]." *Sovietskii Sever* [Soviet North], no. 2: 82–93.

Petri, Bernhard. 1927a. "Okhotnich'i Ugod'Ia i Rasselenie Karagas [Hunting Territories and Residential Distribution of Karagasses]." *Vlast' Truda* [The Power of Labor], 32.

———. 1927b. "Olenevodstvo u Karagas [Reindeer Herding Among Karagases]." *Vlast' Truda* [The Power of Labor], 46.

———. 1928. "Promysly Karagas [Subsistence Strategies of Karagasses]." *Vlast' Truda* [The Power of Labor], 52.

Pomishin, S. B. 1971. "O Transportnom Ispol'zovanii Olenia Tofalarami [On the Transportation Use of Reinder among Tofalars]." *Sovietskaia Etnografia* 5: 128-131.

———. 1981. *Problema Porody i ee Sovershenstvovania v Olenevodstve* [The Problem of Breeds and their Perfection in Reindeer Husbandry]. Iakutsk: Knizhnoe Izdatel'stvo.

Popov, I. I. 1879. "Zapiska o Karagassakh [A note on the Karagas]." In *O Pervykh Priemakh Fizicheskago Vospitaniia Detei u Razdnykh Narodnostei, Naseliaiushikh Irkutskuiu Guberniu* [Methods of Early Childhood Physical Education among Diverse Nationalities Residing within Irkutsk Province]. Irkutsk: Irkutskii Gubernskii Statisticheskii Komitet [Statistical Committee of the Irkutsk Governorate].

Rassadin, Igor. 2000. "Khoziaistvo, Byt i Kul'tura Tofalarov [Household, Subsistence and Culture of Tofalars]." PhD diss., Institute of Mongolian, Buddhist and Tibetan Studies, Ulan-Ude.

Toboliakov, V.N. 1930. "K Verkhov'iam Izcheznuvshei Reki. Sostavleno po Dnevnikam Nachal'nika Gydanskoi Ekspeditsii Akademii Nauk SSSR B. N. Gorodkova. [Into the Upper Reaches of a Disappearing River. Compilations of the Diaries of B. N. Gorodkov, Head of the Gydansk Expedition of the Scientific Academy of the USSR]." *Rabotnik Prosveshchenia* [The Educator], 91–92.

Tugolukov, V. A. 1962. "Vitimo-Olekminskie Evenki: Ocherki Istorii [Vitim-Olekminsk Evenkis: A Historical Overview]." *Trudy Instituta Etnografii. Novaia Seria* [Papers of the Institute of Ethnography. New Series], no. 78: 67–97.

Vainshtein, Sevyan. 1980. "Proiskhozhdenie Saianskikh Olenevodov (Problema Etnogeneza Tuvintsev-Todzhintsev i Tofalarov) [The Origin of the Saian Reindeer Herders (On the Question of Tuvinian-Todhzhinian Ethnogenesis)]." In *Etnogenez Narodov Severa* [Ethnogenesis of the Peoples of the North], 63–87. Moscow: Nauka.

Vasil'ev, V. N. 1910. "Kratkii Ocherk Byta Karagasov [A Brief Study of Karagas Life]." *Etnograficheskoe Obozrenie* [Ethnographic Review], no. 1–2: 46–76.

Vasilevich, G. M, and M. G. Levin. 1951. "Tipy Olenevodstva i ikh Proizkhozhdenie. [Types of Reindeer Herding and their Origins]" *Sovietskaia Etnografia* [Soviet Ethnography] 1: 63-87.

Chapter 5

On the Role of the Horse in Tofa Households

Alex Oehler and Igor Rassadin

The Saian Mountains are considered the home of the world's tallest and strongest breed of reindeer.[1] These animals belong to the Karagas breed of domestic deer, deriving their name from the Indigenous Karagas inhabitants of Tofalaria (known today as Tofas). The Karagas breed is ridden much like a small horse (cf. Ge 2015, 47). In other regions of Siberia, where reindeer breeds are significantly smaller in stature, reindeer are often ridden over their upper thoracic vertebrae to protect their spine (e.g., Vainshtein 1978, 38).[2] In the Saians, Tofa hunters select the biggest of their bulls, riding them near the center of the back between thoracic and lumbar vertebrae (Petri 1927, 10). As a result of their selective breeding, some of the castrated male deer come close in stance and use to the able-bodied small statured Mongolian horse, which is so widespread across Inner Asia.

In spite of the ability of the Tofas and several other Siberian reindeer herding societies to transition—almost seamlessly—between horses and reindeer, relatively little attention has been paid to the relationship between equestrian and reindeer riding cultures of southern Siberia. Many scholars with an interest in the region have focused on reindeer domestication, accentuating the importance of tame *Rangifer* in the households of taiga hunters, focusing on the possible origins of Eurasian reindeer breeding. Part of this may hinge on the fact that deer, possibly even reindeer, feature prominently in ancient petroglyphs of the wider region (Vainshtein 1980, 120). This interpretative emphasis is contrasted by Chinese chronicles of the third century, which report "Dubo, Milige and Echzhy" clans as residing north of Lake Khövsgöl.

These clans are said to reside in bark-covered conical lodges, much like the ancestors of present-day Tofas, and they are described as having "many good horses" (Bichurin 1950, 354).

This chapter combines a review of the ethnographic literature for Tofalaria, the traditional territory of Indigenous Tofa hunter-herders, with ethnographic interviews and observations made during three weeks of fieldwork in 2014 and 2018. It also draws on the long-term fieldwork (1998–1999) of Tofa-Russian historian and ethnographer Igor V. Rassadin, the son of well-known Russian linguist Valentin I. Rassadin. The ethnohistorical materials reviewed here cover nearly 200 years of Tofa history, with the earliest sources being those of explorers[3] who traveled through Tofa territory prior to their settlement into villages. These limited pre-revolutionary accounts precede more systematic post-revolutionary surveys conducted under the newly formed Soviet state.[4] Several of the latter works informed state-organized efforts of settling mobile Tofa hunter-herders in the process of forming collective farms. These latter studies are among the more insightful accounts of mobile Tofa life. Today Tofas reside in three permanent settlements; Alygdzher, Verkhniaia [Upper] Gutara, and Nerkha.

Post-Soviet scholars of Tofa ethnohistory (e.g., Krivonogov 2008; Rassadin 2000; Mel'nikova 1994) have relied on these early twentieth-century accounts to plot more recent changes in material culture and ethnic identities. To contextualize these accounts, along with our own observations, this chapter takes into account some of the literature from the three adjacent Indigenous Saian groups, among whom are Soiots of Okinskii District (henceforth Oka) in western Buriatia, Tozhus in the eastern part of the Republic of Tyva, and Dukhas of northwestern Mongolia.[5] Together, these four Indigenous groups share what Russian archaeologist and ethnohistorian Sevian Vainshtein (1980), along with others,[6] terms the "Saian type" of reindeer herding. In all four cases this type of herding has been accompanied by an ethnographically underreported equestrian culture.

Much of Soviet era scholarly discourse on South Siberian reindeer pastoralism framed the Saian type as a culturally evolutionary stage in what was thought of as a more or less linear progression toward a Netnets-like type of large-scale "reindeer production" (Rus. *olenevodstvo*).[7] Among practices commonly associated with the domestication and productive intensification of herds were their use for transportation, milk production, and saddling techniques. In contrast to this view, we are interested in the deliberate small scale of the Saian type of reindeer pastoralism. Not only do we see it as a self-sufficient adaptation to a highly specific environment, but as a practice that stands in intimate relationship with an ancient and recurring tradition of alpine and subalpine horsemanship.

HISTORICAL OVERVIEW

Finnish scholar Matthias Alexander Castrén visited Tofas on his travels from Kansk to Nizhneudinsk, between November and December of 1848 (1856, 385–95). How much he actually observed of Tofa life at this time of year is unclear. In his letters from Nizhneudinsk he focuses primarily on linguistic parallels between Turkic and Mongolian dialects in an effort to distinguish the ethnic roots of Tofas. Yet, in a later publication (Castrén 1857) he provides a detailed comparative study of Turkic Karagas [Tofa] and Koibal, which indicates significant contact time with speakers of Tofa. Neither of these publications, however, contain ethnographic detail, let alone information about Tofa use of horses.

Castrén was soon followed by Russian statesman Iulii I. Shtubendorf (1858). More ethnographic in essence, Shtubendorf writes about the role of Tofas in locating gold runs on Khorma Ungurbei River. These locations were later turned into productive mines. Bemoaning the fact that the discovery of gold had made Russian prospectors rich while leaving their Tofa guides poor, Shtubendorf does little to describe the role of Russian horses in this mining work. He merely describes the advantages of Tofa reindeer as a means of transportation over horses in mountainous taiga. Shtubendorf mentions that the Tofas he encountered were self-identifying as baptized Christians in the fourth generation. This observation confirms relative proximity to Russian settlers, which may have begun as early as the 1600s.

Around the time of Shtubendorf's visit, another well-known resident of Irkutsk, publisher Ivan I. Popov (1879), reports on his encounters with a Tofa household in possession of reindeer, but with no mention of horses. Popov mentions earlier brief accounts of Tofa life, one written by an obscure German traveler, the other by an Orthodox priest named Prelovskii. Like several others passing through the area, Shtubendorf and Popov seem to have traveled in winter. Tofa horses would have been located at their winter pastures, which explains why neither author makes any mention of Tofa horses, focusing on reindeer instead. No explanation is given for the choice of season for travel, but it can be assumed that the frozen rivers served as more navigable roadways.

About thirty years later, German-Russian Turkologist, Vasilii V. Radlov (Radlov 1907, ii) along with his student and colleague, Khakas-Russian turkologist Nikolai F. Katanov (1891), encountered Tofas on several occasions over the winter period from February 10 to March 21, 1890. Together they recorded Tofa oral traditions while visiting on Kan', Birius', and Uda Rivers. In their notes we find a detailed description of a shaman's horse consecration ritual (Radlov 1907, 608). Both scholars also write about a

Tofa elder who used his horses on extensive trading tours with neighboring peoples. The man preferred horses to reindeer because they better managed the heat and bugs of summer (Radlov 1907, 615).

In a Tofa tale, recorded by the two turkologists, an elder mentions a group of horse-hoofed people[8] pursuing a (Tofa?) traveler as he flees through a swamp, himself riding a horse (Radlov 1907, 654). Finally, in what Radlov calls a "Story About Horses," a Tofa consultant reports: "We do not keep many horses. Some of the horses we have are white, red [Rus. *ryzhaia*], blue, black [Rus. *voronaia*], light brown [Rus. *savrasaia*], red with a white mane and tail [Rus. *irgenevaia*], light yellow, or golden [Rus. *solovaia*], dark brown, descended from red [Rus. *buraia*] and unevenly patterned ones [Rus. *pegaia*]" (Radlov 1907, 611).

Eighteen years later, in 1908, Russian ethnographer and folklorist Viktor N. Vasilev (1910) spent two months with Tofas, as part of an anthropological expedition, commissioned by the Museum of Anthropology and Ethnography at Saint Petersburg. During his visit he produced remarkable photographs, some of which appear in the 1910 report consulted for this chapter. In spite of the summery scenes depicted in the photographs, none show horses. Vasilev notes, however, "If someone has 40–50 reindeer, a horse and a complete conical tent, such a person is already rich" (1910, 20).

Confirming the observations of Radlov and Katanov, Vasilev reports several Tofas as having two or four horses. He notes, "Karagas [Tofas] look after their horses as they do after reindeer—that is, they keep them under a self-provisioning regime year round, which is why every year by spring time a significant part of them die from hunger, and they are necessitated to replenish them by purchasing new ones" (Vasilev 1910, 8). Vasilev does not explain where his interlocutors obtained their horses annually, but given the small number of horses per household, it would seem that they were not engaged in major breeding of their own.

TOFA HORSES IN THE ANNUAL CYCLE

Historical accounts, including more recent ethnography, speak of Tofa reindeer being used from first snow in late autumn until the pelt hunt season in early spring. At the end of spring, as temperatures rise, the reindeer are released to higher altitudes where they are left to graze all summer, accumulating fat for the winter season. Even at the time of our field visits, it was considered disadvantageous to put reindeer to work during their summer resting period. To fill the seasonal void in transportation, Tofas employed horses.

During I. V. Rassadin's (2005, 34) fieldwork, between 1998 and 1999, Tofa elder consultants V. V. Adamova, G. I. Adamova, S. N. Araktaev, and N. E. Kangaraev reported that horses had been used by Tofas in the past, and that they had used Buriat-made saddles, which were interchangeable between reindeer and horses. Buriat saddles were designed to be used on Mongolian horses, and because Tofa horse groups had been imported from Okinskii District, these saddles fit them well. According to the elders interviewed by I. V. Rassadin, Mongolian horses from Oka stood only a little taller than the Karagass breed of reindeer.

It follows that the near equal height of these two species reduced the number of saddle types required in each household, even if households did possess more than one type of saddle.[9] The wooden saddle design, brought to Tofalaria from Oka, consisted of four pieces (front arch, rear arch, right and left bars, see also Tkačenko 2010, 3) of hand-shaped wood, fitted together like a puzzle and held taught by strips of leather. These riding saddles were largely interchangeable between small Mongolian-type horses and full grown trained Karagas reindeer, and they could still be found in both Tofa and Soiot households during our fieldwork.

The wooden horse pack and riding saddles had become less common in recent years, as they were no longer being produced in Oka. Instead, factory-made horse saddles, based on two steel tube arches, had become the most common saddle design in both regions. Tofas and Soiots would adapt (or "lighten" [Rus. *oblekhchennoe sedlo*]) these newer designs to fit their Karagas breed reindeer, retaining the interchangeability between horses and reindeer. Both wooden and steel tube designs were covered with reindeer or bear fur as padding for the rider, and the frame itself was underlain with felted wool blankets to protect the animal's spine.

The contemporary lightened saddle consisted of right and left wooden bars, bolted to two steel arches (see Figure 5.1). A strong 10 centimeters wide leather strip stretched across the vacuity between the bars. Connecting the arches, the leather strip was tied to both wooden bars in multiple locations. Felt, foam, or leather overlay this construction underneath the rider's saddle fur. The modular construction of both styles, their ability to be repaired or improvised, and their relative durability and structural flexibility allowed the construction to be passed down from one generation to the next. The possibility to add or subtract bulk from the saddle also allowed transitioning its use from winter to summer seasons, and from reindeer to horses.

Old-style saddles had been made from well dried birch or larch wood (Mel'nikova 1994, 54). The front and back arches were usually decorated with carvings, red painting, or embossed metal plaques (Mel'nikova 1994, 54). Rather than making their own saddles, many of the designs still found in

Figure 5.1 Lightened Horse Saddle for Riding Reindeer, Oka. *Source*: 2013, A. Oehler.

Tofalaria today originated from Oka and Todzha (Eastern Tyva). B. E. Petri (in Mel'nikova 1994, 55–57), and later Vainshtein (1974, 165), analyzed their designs to map possible artistic connections between Tozhus of Eastern Tyva and Tofas. Buriat saddle makers decorated the wooden front and back arches of their saddles according to their own symbolic traditions, but Tofas on occasion would adapt these designs to fit their cultural context.

We encountered an example of this during our fieldwork in 2014. At the autumn reindeer camp on Dugul'ma River, resting beneath a tree, along with various other implements, we found a red wooden riding saddle in the Mongolian four-piece design. It had been redecorated with an embossed metal plaque displaying a simplified Russian Orthodox icon (see Figure 5.2). The saddle seemed to be stored for winter service, paired with a working chainsaw, awaiting its owner who was at the time collecting his transport deer. The combination of red paint with the Virgin Mother holding her Christ Child is representative not only of the intersecting of Buriat Buddhism with Tofa Orthodox Christianity, but also of a possibly much older mixing of equestrian and reindeer riding cultures.

The purpose of horses, in this mixture of traditions, would primarily have been to allow reindeer a period of recovery. But the fact that winters were filled with strenuous travel relating to hunting activities did not mean summers were necessarily a resting period for Tofas and their horses. In summer, people attended weddings, conducted various festivities, traveled to Oka to attend

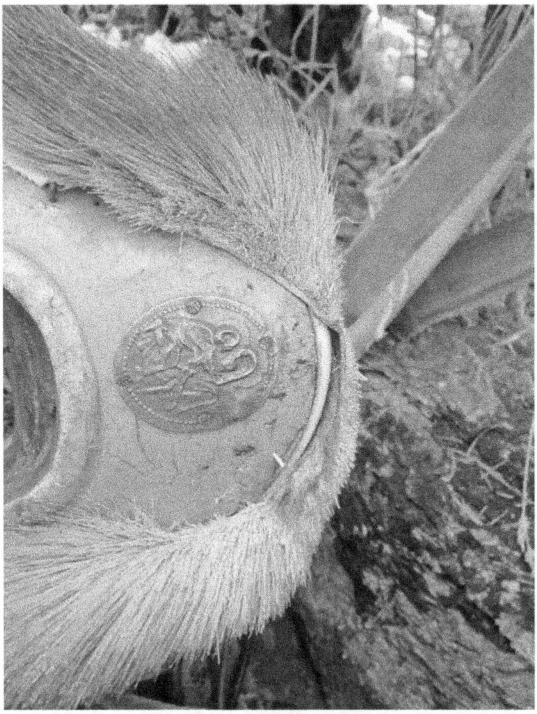

Figure 5.2 Horse Saddle Used for Reindeer with Icon, Tofalaria. *Source*: 2014, A. Oehler.

the Buddhist temple, visited distant relatives, and traded with Russian settlers (Petri 1927).[10] To transport the amount of cargo required for these engagements would often have required the help of horses. A horse could carry loads of 65 to 82 kilograms over long distances (I. V. Rassadin 2003, 35). In so doing, one horse could replace up to three reindeer in carrying capacity.

Given this advantage, horses were also used to bring winter supplies of flour, oats, salt, and ammunition into the taiga. On average it might take one trip with three horses to stock a hunter's winter camp supplies before the first snowfall. The supplies were then stored in raised cash platforms at autumn and spring camp locations, from where they could later be gathered and further distributed with the aid of reindeer (I. V. Rassadin 2005, 35). This strategy was still being used during our fieldwork, although a number of herders were increasingly relying on motorized vehicles, such as four wheelers, depending on the accessibility of their hunting sites. In spite of the greater weight-bearing capacity of horses compared to reindeer, it was primarily the horse's inability to locate sufficient grass in dense taiga that restricted its employment in deeper forested settings.

Some taiga hunters, such as the Tungus-speaking Oroqen of northeast China, are said to have trained their horses to travel on extensive trails without intermittent glades. Unlike their reindeer-herding relatives on the Siberian side of the Amur River, Oroqen hunters on the Chinese side kept only horses (e.g., Noll and Shi 2009, 123). Elizaveta N. Shirokogorova (1919, 30), spouse of famed Russian anthropologist Sergei M. Shirokogorov describes, in her field notes from northwestern Manchuria, how Oroqen riders trained their horses to subsist on a raw meat diet for weeks on end. This allowed hunters to venture with their horses into territory normally accessible only to those using reindeer for transportation. Such a tradition is not mentioned in the literature of the Eastern Saian Mountains. It follows that the Tofas made use of glades along the extensive trail system that connects their region to Tozhu, Buriat, and Soiot territories. During our fieldwork, this trail system was not as commonly used as it once had been when Soiots, Tofas, and Tozhus had visited each other much more regularly.

While planning a multi-day horse trek to a Tofa reindeer camp in the autumn of 2014, our Tofa guides interpreted the feasibility of multiple routes in relation to known distances between accessible glades. These glades served as overnight campsites, allowing the hobbled horses to selfprovision during the night in preparation of the next day's trek. Reindeer treks do not require this type of vegetation planning, as the deer can feed on lichen found above a certain elevation throughout the taiga, as well as on tree lichen. As a result, some historical trails in the Eastern Saians are known specifically as former reindeer trails, while others have been shared between reindeer and horses. Abandoned reindeer trails can still be identified by the type of groove left in the ground as a result of the traditional reindeer caravan harnessing system (see Pomishin 1971, 129), which was still being used by hunters during our visits.

Swiss-Russian anthropologist and archaeologist Bernhard E. Petri, who conducted extensive research in Tofalaria in 1926, reports Tofas annually visiting the Buddhist temple in Oka during the summer (1927, 35). This is a curious piece of information as Tofas, like their Soiot and Tozhu neighbors, had their own shamans who were often marginalized by itinerant Buddhist lamas. In contrast to Tozhus and Soiots, being under the influence of settlers who converted to Buddhism, Tofas were actively proselytized by the Russian Orthodox Church of Irkutsk. As mentioned previously, many converted to Christianity (e.g., Mel'nikova 1994, 109). It is likely that Petri is referring to Zhelgenski *datsan*, the Buddhist temple that opened near Balakta in 1906 (Sharastepanov 2008, 35). Since these travels took place in summer, Tofas would have taken their horses on this trip.

It would have been during these visits to Oka that Tofas traded horses they had previously obtained from Russian settlers in Tofalaria with their

Buriat neighbors in exchange for Mongolian horses (Petri 1927, 36). Our Tofa consultants, on several occasions, confirmed the relative uselessness of Russian horses in the taiga. Not only were they picky eaters, their stature and lack of agility were counterproductive in steep and difficult terrain. Russian horses easily broke their ankles, and they rarely survived two years of service in mountainous terrain. It made sense then to exchange such horses for the smaller more agile and robust Mongolian breed.

Conversely, according to Petri, Oka Buriats were interested in the Russian breed because they were able to pull carts, unlike Mongolian horses. Historically there has not been much agricultural development in the Oka Valley, but likely it was their extensive hay harvesting and manure distribution in pastures that called for the use of carts.[11] It would have been during their summer trading visits to Oka, that Tofas obtained wooden riding saddles from their Buriat neighbors. Further research is needed to show how Tofa and Soiot reindeer pack and infant-carrying saddles resembled each other, and whether there occurred trade in these items between the four reindeer-breeding societies or not.

Beginning in autumn, when Tofas transitioned back to the use of reindeer, they would release their horses into winter grazing grounds. Here the horses were able to move without restriction. They were not hobbled, fenced, or in any way supervised. Prior to Soviet settlement, the winter grazing grounds had been clearings in the taiga. According to local memory (I. V. Rassadin 2005), no one would steal horses from these locations, and neither was there any significant pressure from predators. Since the time of Tofa settlement in villages, horses had been self-provisioning within the much smaller radius of the villages. At Alygdzher, the community's main winter glades were located along the Uda River, just upstream from the settlement. Crusty snow could form in these exposed areas in late spring, which periodically led to serious malnutrition, and sometimes resulted in death. In 2018, a number of Tofa families maintained hay harvesting claims along the river from where additional feed was boated to the village. Primarily, these provisions were intended for dairy cattle during the winter, but on occasion the hay was also used to subsidize horse feed.

It is unclear from the literature where the winter horse glades of pre-sedentary Tofas were located. We do know, however, that Tofa winter horse groups consisted of animals belonging to several related households. Two or three people were selected to drive the horses of multiple families to their joint winter range in autumn, from where they would be collected again in spring. Following the winter, the horses had become somewhat wild, requiring animals and riders to become re-attuned to each other (I. V. Rassadin 2005, 36). Interestingly, the prolonged separation from humans did not affect reindeer in the same way it had affected horses (see K. B. Klokov, chapter 5, this volume). Following Tofa settlement into villages, one of the first activities after the spring roundup was the transportation of firewood from the taiga back to the village. This strenuous activity served as an annual reaffirmation of the horse-human bond in spring. More recently, motorized vehicles have largely replaced the role of horses in firewood transportation.

The question of where exactly Tofas once held their horses, and to what extent horses intersected with reindeer has not been answered yet. In an effort to develop new methods to establish the presence of animals in contemporary and ancient environments, we collected fecal deposits with our archaeologist and palynologist colleagues (see Harrault et al. 2019, 3) in 2014. Being able to distinguish between different species on the basis of their fecal matter in ancient deposits would help establish a record of the presence of horses in pre-contact Tofa reindeer encampments. It would also allow archaeologists to learn more about ancient seasonal uses of grazing grounds throughout the Eastern Saians, including the presence of other species in Tofa households. At this point, fecal matter analysis can only distinguish between omnivores and herbivores, but not between species.

ON THE COSMOLOGICAL SIGNIFICANCE OF HORSES

The ethnographic record for the Saian-Altai region contains multiple references to the cosmological significance of reindeer. Less common are references to the role of horses in reindeer-dominated economies. B. E. Petri's Tofa account of 1927 is an exception (1927, 38–39) in this regard. He explains, how under the guidance of a shaman, horses were dedicated to the mountain spirit master *dag eezi,* much like reindeer. This could be done in response to an illness, for success in one's hunt, in case of one's upcoming marriage, for the formation of a new household, or in supplication of the mountain spirit's general benevolence toward one's family. For this purpose, a shaman was summoned to ascertain the type of horse desired by the spirit master. A householder would go out of his way to locate the desired type, searching not only among the horse groups of his relatives, but also among those held by Buriats in Oka and among Russian settlers.

Once the household head had located the desired animal, he would purchase it at any cost. Upon the householder's return, the shaman would tie colorful ribbons to the mane and tail of the animal. The horse would then be released to join the household's group of horses. Only the male head of the household was now allowed to ride the sacred animal, while women and girls were not allowed to be seated on it. Similarly, it was considered a taboo, according to Petri (1927, 39), to saddle the consecrated horse with a saddle on which a woman had previously sat.

Horse consecration closely resembled similar procedures formerly practiced on reindeer (Petri 1927, 32). Both horse and reindeer consecration were based on the belief that humans and animals lived on the land of the spirit master dag eezi. Consequently, all life forms depended on the benevolence of this master. Animals consecrated to the spirit master were more likely to enjoy its protection than animals that belonged to human households. However, a herd belonging to a human household, among whom resided a consecrated animal, was thought to benefit by extension. It made sense then to have consecrated animals among one's herds, whether or not the selected species were native to the land. The fact that horses were occasionally selected by shamans for consecration, suggests that the spirit master had a liking for animals not present among its own herds.[12] Petri does not indicate in his description of horse and reindeer consecration rituals, which of the two was considered older. At the time of our field visits, neither of the rituals seemed to be practiced anymore. However, anthropologist Selcen Küçüküstel (2018, 166 and chapter 9, this volume) describes a near identical practice she observed during her work with Dukha reindeer herders of northern Mongolia.

The oral traditions of Tofas suggest horses may have played a significant role even prior to the arrival of Russian settlers. This seems evident from Tofa folklore and myth, where heroes are found riding not only reindeer, but also horses. Examples abound in the sayings collected by V. V. Radlov (1907), as well as in the Tofa stories and songs collected more recently by Russian linguist V. I. Rassadin (1996). A good example comes from the origin account for the Orion constellation, in which Tofa is called *ysh myigak* (lit. three female red deer). A man was hunting three red deer with the aid of his two dogs, while sitting on a horse. The hunter had bred the horse and dogs on his own for the purpose of hunting. However, *burkhan* (the master deity) lifted the whole group up into the sky, where together they became the constellation of the "three female red deer" (V. I. Rassadin 1996, 8–9).

TOFA HORSE FOLKLORE

Russian linguist V. I. Rassadin's goal was to assess how much the Tofa language had changed since turkologist Nikolai F. Katanov's first recordings of their oral culture in 1890 (published in Radlov 1907). In his compilation of old and newly recorded folklore, V. I. Rassadin combined his own recordings with those of others, including the collections of A. K. Stoianov (1980). Horses are mentioned on multiple occasions, including in heroic tales, songs, and riddles. In one story, folk hero Mikhishka observes from a tall cliff how his entire household packs up and leaves for their summer pastures at higher elevation. The household migrates with "all its deer and horses" (V.

I. Rassadin 1996). It can be inferred from this story that Mikhishka's mobile household took their horses to the summer reindeer pastures after they had been gathered from their winter glades. Even today, this remains a common sight, particularly when hunters arrive on their horses to select riding deer for the winter (see chapters 5 and 7, this volume).

In another tale, orphan hero Eskius-ool arrives with nine reindeer at an unrelated family's encampment. He notices a single horse with an injured spine among his host's animals. The family tries to convince the hero to release his deer for grazing among their herd and to spend the night at their encampment. Eskius-ool refuses, saying the host's horse will eat his deer. After a while, however, he relents, and having fallen asleep, the family secretly butcher all his deer. They hide the meat inside the saddle packs of their own transportation reindeer. Eskius-ool is left with the injured horse. Although his own sins will revisit him later, he rides the horse to a future in which he establishes a family of his own (V. I. Rassadin, "The scapula bone," recorded at Alygdzher, 1965, from I. S. Ungushtaeva).

The fact that Eskius-ool's new horse has an injured spine may suggest its Russian origin. The absurdity of the horse eating the reindeer is paralleled with a previous mention of a scapula bone being eaten by another host's cooking pot. Both instances seem to be metaphorical of the predatory intentions of the hosts and their *aal* (household). In either case, the second hosts perform an unusual task in butchering the hero's reindeer. Under normal circumstances reindeer were not used for their meat, unless they had been injured beyond recovery. Instead of slaughtering the injured horse, Eskius-ool is given it as compensation. The horse, in this case, becomes foundational in the establishment of Eskius-ool's own aal.

The theme of the horse and its injured spine recurs in another variant of the scapula bone story. This time, a nameless hero and his ailing wife live in poverty and without animal herds. After the woman dies from illness, the man rides off on his injured horse. He shrewdly accuses the members of another household of the spinal injury in his own horse. The man obtains compensation from the accused household in the form of a healthy horse. Shortly after, his wounded horse dies in the meadows of the accused household and the man pulls a grey scapula from the open wound of the animal's carcass.[13] He trades the bone for ten wethers from the same household. (V. I. Rassadin 1996, "Gray Scapula," recorded at Alygdzher, 1972, from N. S. Bakanaev). In both variants of the story, a poor individual uses a single injured horse as the foundation upon which he builds a multispecies aal through trade and trickery.

Not all stories position horses in impoverished households. Borrowed from Oka, the story of "Uluzun-Khan and the wise girl" tells of a powerful man who challenges a young woman's younger brother to ascertain the age of a bear and later of a hare. The elder sister's advice ensures the boy succeeds each time. Finally, the Khan says, "Tomorrow you will train a black stallion with nine stirrups." That night, the boy's elder sister makes him three iron whips. Having mounted the stallion on the following day, the boy uses one of his whips. As he sits down on the horse, the stallion begins to resist. The boy keeps striking the animal until all of his whips have been used. At this point the horse cries out, "-O -o!" As the boy is looking on, he realizes the horse has turned into the Khan himself (Katanov, "Uluzun-Khan and the wise girl," from Nikolai I. Shyryshpanov, re-recorded by Rassadin 1968).

In the story, the Khan verifies the boy's report of the animals' ages by looking into his *sutra*, his "magical book" or "book of fate" (V. I. Rassadin's 1996 translation). The mention of a sutra[14] makes sense here, as the story is said to be a loan from Oka where Buddhism became popular among Buriat settlers, as well as among some Soiots, beginning in the late nineteenth and early twentieth centuries (Sharastepanov 2008, 33). Oka-Buriat tantric lamas to this day divine the age of their clients. The transformation of the Khan into a stallion, as well as knowledge of the allotted life span of individual animals, more resembles pre-Buddhist accounts of spirit masters. Spirit masters would often take on the form of animals belonging to their herds (e.g., Donahoe 2012, 102–103). In either case, the story exemplifies the glory of a black stallion by association with a powerful master.

The song, "When riding I was, girl of Kara-Chogdy,"[15] re-recorded by V. I. Rassadin (1996) from the words of N. S. Bakanaeva at Alygdzher in 1972, indicates a variety of horse coat colors and breeds known in Tofa folklore: "When riding I was, girl of Kara-Chogdy / No dark bay was following me.[16] When riding I was, girl of Chogdy / No racehorse was following me. When riding I was, girl of Khash / No palomino was following me. When riding I was, girl of Sary-Khash / No small statured roan was following me. When riding I was, girl of Cheptei / No two year old trotter was following me." It is not clear whether the combinations of coat color or breed and the girls' listed clan affiliations are significant. Neither is it evident why the horses in the lyrics do not follow the young woman (or women). Yet it speaks of a variety of horse breeds, the last among whom likely refers to a Russia Orlov Trotter.

Another coat color is mentioned in the following riddle: "Inside a hairy fence; a pinto horse. [What is it?]—[It is] the eye" (Rassadin 1996, riddle 24). However, the most preferred coat color seems to have been the yellow whitish palomino. Yellow-white or yellow-brown horses appear repeatedly in Tofa folklore. In the song, "The fullest among seven girls," the palomino is compared to a lovely young woman. N. F. Katanov recorded this song in 1890 from Evdokia Khodogonova, and it was still known in 1974, when V. I. Rassadin recorded it again from the words of N. S. Bakanaeva at Alygdzher: "The fullest among seven girls / Is she not my girl?! The one with the greatest mane among seven horses / Does it not belong to me?! The loveliest among six girls / Is she not my girl? That palomino among six horses / Does it not belong to me?!" Here preference for coat color is given, while other songs indicate preferred stature, size, and type of mane—again in relation to preferences for specific human attributes found in the opposite sex.

Thus, in the song, "The middle one among ten horses," a young woman sings: "The middle [Rus. *srednii*] one among ten horses—it ought to be my horse! The middle one of ten boys—he ought to be my friend!" A young man responds: "The one with the greatest mane among seven horses—it ought to be my horse! The most mature [Rus. *vzroslaia*] among seven girls—she ought to be my friend!" (Recorded from the words of E. M. Amastaeva by A. K. Stoianov at Alygdzher, 1976, in V. I. Rassadin 1996). Interestingly, not the tallest or strongest of horses or men are given preference, but the most average. Perhaps this has to do with the ideal balance between endurance and agility in a mountain environment. The young man, by contrast, seeks a great mane, paired with maturity (in the sense of being an adult). Maturity likely speaks to the ability to procreate and astutely manage an ever increasing aal.

The close relation between horses and reindeer in Tofa households is further evidenced in the poetic "Response of the master spirit of Nostol taiga to a Shaman's spell regarding illness" (V. I. Rassadin 1996 [from1972 field notes]). It is a recording of the words of the spirit master in response to the shaman who has been commissioned by a householder to conduct the animal consecration ceremony mentioned previously: "Golden nightingale [Rus. *zolotisto-solovyi*] horse / If it will be consecrated. A three year old white reindeer / If it will be consecrated. In the great life of children / Illness and life. They equate! Great illness and whitish fog. They disappear!" (V. I. Rassadin, (1996) [from 1972 field notes], recorded from the words of N. S. Bakanaev at Alygdzher). Here a golden and pale yellow horse with light tail and mane is equated to a three-year-old reindeer of white[17] coat color, either of which are found pleasing to the spirit.

Radlov (1907, 608) provides more detail on the actual logistics of the horse consecration ritual. What he does not go into detail on is the shaman's nighttime rite, which followed the morning ritual. Likely, the scholar was not present for the shaman's interaction with the spirit. It is also possible that the consultants themselves were not present, or that they were not comfortable speaking to strangers about such things:

> When a horse must be consecrated, first [people] gather in the yurt, while the horse stands outside the front door. A white felt [*koshma*] is spread out beneath the horse[18], as the horse faces east, toward the rising sun. After tea has been had, the vessel holding the tea is carried outside, where it is sat down on the white felt. The tea, its vessel and seven tea cups are placed on the white felt to the left side of the horse. The seven tea cups are all located in one place. Scooping tea

> from the seven cups with a spoon, the tea is poured over the horse: on the head; spine, legs and tail. Having poured the tea [over the horse][19], the shaman washes the horse with his own hands. The consecration of a horse takes place in the morning, at sunrise. The shaman does not perform his rite here; he does so only at night. Having washed the horse, the shaman ties cut strips of white, red, blue, black, and yellow cotton cloth [*rubakhu* (i.e. *loskut' cittsa*) to the horse's tail and mane. It is also tied around the neck. Standing to the left side of the horse, the shaman burns wild thyme [Rus. *bogorodskuiu travoiu* (Thymus Serpyllum L.) as incense. Those present walk about, looking on in silence. The consecration of animals is conducted to ensure people do not fall sick and reindeer do not go missing or fall ill. No one, except the master [Rus. *khoziain*], will sit on the ydyk. If anyone else sits on the animal, the horse or reindeer, or even the person who sat on it, will be lost. [. . .] The ydyk is consecrated to the spirit master of the mountains and waters.

What we see here is an example of the flexibility of Inner Asian ritual practices that can be extended and adapted to accommodate diverse species. In this way the ritual practice resembles the material culture, which in the form of multi modular tack is equally interchangeable between diverse animals.

DISCUSSION

This chapter set out with the argument that horses may not be a late addition (or recent innovation) to Tofa households, but that, along with reindeer, they represent an ancient self-sufficient adaptation to challenging subalpine and alpine environments. As such, the presence of horses in Tofa households may be an extension of an ancient and recurring Turkic tradition of horsemanship in the Eastern Saians. Whether or not this argument will stand depends in part on future archaeological research. To date, no significant excavation has taken place in Oka or Tofalaria, and existing ethnohistorical data suggest that the role of horses in taiga hunting households differed from that of ancient Turks in the more westerly plains.

We know, from adjacent Tyva, of the significance of horses in the afterlife of ancient Turks. Horses were often buried next to their deceased riders, inside mound burials. Russian archaeologist, ethnographer, and ethnohistorian Sevyan Vainshtein observed the continuation of this practice in contemporary Tyvinian culture of the first half of the twentieth century (1957, 185). The close tie between humans and horses in that cultural context is further attested to in that a Tyvinian hero had no name until he had acquired a horse for himself (Vainshtein 1969, 125–126). In Tofa heroic folklore, as we have seen, the role of horses and reindeer seem to have been more interchangeable, and at present there is no evidence of joint burial.

It is difficult to say, with any certainty, when Tofa horsemanship became integral to the mountainous households of Tofalaria. Even so, there exists linguistic evidence suggesting a tradition of horsemanship predating both earlier imperial and later Soviet forays of Russians into Tofa territory. The ancestral Tofa language, which is now moribund, contains largely unaltered Turkic (i.e., Shor, Koibal, Kachin, see Vainshtein 1980, 71) horse terminology. V. I. Rassadin (1971, 73–74) has identified terms, such as the generic *at* for horse, *askyr* for stallion, *akta at* for gelding, *be* for mare, *hylyn* for a foal of up to one year, and *chapaa* for a foal of two years. Included are terms for the material culture associated with horses, such as *ezer* for saddle, *kuburha* for the crupper connecting saddle to tail, *hoemyndyryk* for the breast collar, *yngyrshak* for the pack saddle, *kymshy* for the whip, *tushakh* for front hobbles, *taha* for a horseshoe, *chep* or *chel bakh* for rope made of horsehair, and *hyl* for hair from a horse tail.

Although many Tofa households still own smaller numbers of horses, local opinions regarding the history of horses in Tofalaria vary. For example, in 2018 reindeer herder Sergei V. Tulaev was of the opinion that Tofas never really had horses of their own, and that they were a Russian introduction. In this view, horses were a byproduct of gradual assimilation to Russian settler culture, a process often described by Tofas in the phrases, "we have been Russified," or "we have undergone métissage" (Rus. *my obrusilis* or *my obmetisilis*). According to Tulaev, horses first appeared with the emergence of the Russian border guard, which for several years was located at Alygdzher. The guard had patrolled the border with Tyva, which was recognized as a nation independent from Mongolia and the Soviet Union between 1921 and 1944. Also, Russian geological prospectors brought in horses to transport out mineral samples, using horse-drawn sleighs.

In spite of the importance of horses in the summer activities of Tofas, early oral accounts and subsequent survey records report relatively small numbers of horses. In 1914, Russian geographer and statistician, K. N. Mirotvortsev, counted no more than 56 horses belonging to 98 households (quoted in Petri 1927, 40–41). A decade later, in 1925, Petri counted 130 horses across a total of eighty households. We know from Petri's expedition that Tofa reindeer herds experienced a massive decimation resulting from an outbreak of epizootic disease in 1919. Following this disaster, Petri reports Tofas purchasing reindeer and horses from their neighbors in Oka. Obtaining both species in Oka suggests Tofas were likely dealing not only with Buriat horse breeders, but also with Soiot reindeer breeders.

While Tofas have an extensive reindeer dairying tradition, the Turkic tradition of mare milking does not seem to be referenced anywhere for Tofas. Their Tyvan neighbors, by contrast, milked their mares several times per day from June to September, obtaining approximately 2.5 liters of milk per day

(Vainshtein 1980, 66). The absence of such a practice among Tofas likely has to do with the small number of horses kept on average, and the fact that they were not actively being bred, thus lacking a reliable milking regime as was in place for reindeer. Beginning in spring, Tofa reindeer cows and calves were kept in such a way as to maximize milking opportunities. After the calving period, lactating cows were tied to vertical posts, with their calves tied to movable horizontal logs. In this way, 200 to 300 milliliters of milk was collected per day per cow when milked twice daily (Mel'nikova 1994, 70). There does not seem to be evidence of a similar practice for horses.

The small number of horses in pre- and post-revolutionary Tofa households could be interpreted as indicator of the animal's secondary role, in much the same way that Soviet scholars understood Tofa small scale reindeer breeding. Conversely, it can be argued that horses have fulfilled an essential element in Tofa annual migratory regime. It is the horse, after all, that continues to enable regular resting periods for reindeer, and without horses Tofas may not have been able to adapt to the seasonal and elevation variations of their environment as well as they have. At the very least, the horse has helped Tofas maximize their adaptive capacity to changing social, ecological, and epizootic circumstances. Reliance on horses increased whenever reindeer herds were suffering from diseases. Horses also allowed Tofas to maintain social relations with their Soiot, Buriat, and Tozhu neighbors at the only time of the year unoccupied by their primary mode of subsistence—taiga hunting.

CONCLUSION

It is appropriate that horses would play an integral part in the transportation system of a people with access to the largest reindeer in the world. The two species' unique capabilities not only complement each other, but their comparable size enables partial cross application of tack. It is all the more surprising that in spite of multiple ethnographic references to horses in reindeer-breeding households, relatively little attention has been paid to the relationship between equestrian and reindeer riding cultures in South Siberia. In line with our findings in early Tofa ethnography, as well as our own ethnographic observations, this chapter has argued for the possibility of an ancient tradition of small scale alpine horsemanship in the Eastern Saians.

From Castrén to Shtubendorf, and from Radlov and Katanov to Vasilev, the ethnographic record for Tofas suggests a form of horsemanship that has been intimately interwoven with reindeer breeding. The long-term accommodation of both species is evidenced in the interchangeability of modular saddle designs, as well as in the flexible nature of Inner Asian religious understandings that are able to incorporate diverse species, including animals that lack a direct counterpart in the wild. Furthermore, detailed knowledge of the location of taiga glades has resulted in the strategic placement of horse trails, which in turn enabled Tofa brokerage between Buriat and Russian settlers, including the trade of specialized horse breeds.

The cosmological significance of the horse is perhaps best expressed in its equal standing with the reindeer in shamanic consecration rituals. Here the landscape's spirit master will choose a horse for itself in spite of the absence of horses within its own holdings. The diversity of known horse types in Tofa folklore, along with an ancient Turkic lexicon of horse terminology, suggests the horse predates Russian colonization in these hunter-herder households. Although the presence of horses may have fluctuated, there are a number of grounds that support the existence of an enduring heritage of small scale horse keeping that diverges in character from adjacent ancient Turkic horse-breeding traditions.

NOTES

1. The ethnographic research conducted for this chapter was made possible with the support of the Arctic Domus project (ERC Advanced Grant 295458) at the University of Aberdeen, UK, the Angus Pelham Burn Award for Northern Research, and Seed Grant (No. 6005720) from the University of Northern British Columbia, Canada.

2. In contrast to Tofa reindeer, Anthropologist Tang Ge (2015, 47), who has conducted extensive fieldwork with Ewenki reindeer herders of northern China, reports: "A reindeer's back is flatter than that of a horse, so a reindeer saddle differs in shape from a horse saddle. Because reindeer are smaller than horses and cannot carry as much weight, only women and children can ride them. Men move around by foot."

3. See, for example, Vasilev 1910; Katanov 1891; Popov 1879; Shtubendorf 1858.

4. See for example, Petri 1926; 1927a; 1927b; 1927c; 1928; Zolotarev 1926; Dyrenkova 1940; Sergeev 1940; Pomishin 1971.

5. See I. V. Rassadin 2012; Pavlinskaia 2002 for Oka-Soiots, Donahoe 2012; 2004; Stépanoff 2011; 2012; Olsen 1915a, 1915b for Tozhus, and Endres 2014; Küçüküstel 2013; Wheeler 2001 for Dukhas.

6. See Vasilevich and Levin 1951, 76–7; Maksimov 1928, 9, 27–28.

7. See Milek, Klokov, Kamerling, Oehler, Anderson, manuscript in preparation, "Identification and Interpretation of Small-Scale Nomadic Herding Sites in Woodland Environments."

8. Katanov (1891) relates a Tofa story about three *kubulgats* (persons able to transform into human and animal forms). The younger brother of a Tofa *kubulgat* apprentice turns himself into a horse.

9. This observation is not unique to the mountainous taiga of the Eastern Saians. V. I. Rassadin and S. M. Trofimova (2018, 162) point out that in Mongolian-speaking contexts horses, camels, reindeer, and yak often used the same saddle.

10. The main fur trade, as well as payment of taxes [Rus. *iasak*], took place annually at their administrative gathering [Bur., Evenk, Tof. *suglan*] in December. The event would move from year to year, and Tofas arrived on their reindeer, as can be seen in historical photographs (e.g., Mel'nikova 1994, 108–111).

11. At the time of Oehler's doctoral research in Oka (2012–2014) Buriats and Soiots used specially trained yak bulls to draw heavy manure carts.

12. This observation echoes other recorded exchanges between human households and landscape master spirits. Soiots of Oka emphasized the importance of exchanging goods otherwise unavailable to the spirit (e.g., processed foods) (cf. Pavlinskaia 2002).

13. Although not mentioned in the story, the scapula, especially of sheep, is often used for divination in the Saians.

14. Sutra generally refers to Buddhist canonical scriptures. In Oka, the term sutra is also used for written out "prayers" that are recited by a lama in a householder's domain to ensure good fortune.

15. Kara-Chogdy, Chogdy, Khash, Sary-Khash, and Cheptei are listed by V. I. Rassadin as Tofa clan names.

16. V. I. Rassadin's Russian translations of the Tofa coat color terms are as follows: *temno-gnedoi* [dark bay], *begunets* [race horse], *zhelto-solovyi* [palomino], *savrasyi* [roan], *rysak* [trotter].

17. White reindeer, including other white animals, were also considered sacred by Tozhu and Dukha hunters and herders (e.g., Donahoe 2012, 105; Wheeler 2001).

18. The spreading of a white felt is an ancient and widely practiced method for the consecration or inauguration of persons and animals in Inner Asia. For an example of the use of the white felt in the inauguration of Buriat khans of the nineteenth to twentieth centuries, see Dzhurabaev 2013.

19. It is common in the Saians and beyond, for rituals involving tea, that the attendees begin by drinking the tea, the remainder of which is then used in the ritual. In Soiot hunting contexts tea is made by the eldest hunter and drunk by all attendees, before it is poured on a burning offering made in supplication to the spirit master by the eldest hunter.

REFERENCES

Bichurin, N. Ia. 1950. "Sobranie Svedenii o Narodakh Obitavshikh v Sredney Azii v Drevnie Vremena [Collection of Information about Peoples that Lived in Central Asia in Ancient Times]." 2: 334.

Castrén, Alexander M. 1856. *Reiseberichte und Briefe aus den Jahren 1845–1849.* Saint Petersburg: Anton Schiefer.

———. 1857. *Versuch einer Koibalischen und Karagassischen Sprachlehre.* Saint Petersburg: Anton Schiefer.

Donahoe, Brian. 2012. "'Trust' or 'Domination'? Divergent Perceptions of Property in Animals among the Tozhu and the Tofa of South Siberia." In *Who Owns the Stock? Collective and Multiple Property Rights in Animals*, edited by A. M. Khazanov and G. Schlee, 99–120. New York: Berghahn Books.

Ge, Tang. 2015. "In the Forest Pastures of the Reindeer." In *Reclaiming the Forest: Ewenki Reindeer Herders of Aoluguya*, edited by Å Kolås and Y. Xie, 45–60. New York: Berghahn Books.

Harrault, Loïc, Karen Milek, Emilie Jardé, Laurent Jeanneau, Morgane Derrien, and David G. Anderson. 2019. "Faecal Biomarkers can Distinguish Specific Mammalian Species in Modern and Past Environments." *PLOS ONE* 14 (2): e0211119. doi:10.1371/journal.pone.0211119.

Katanov, N. F. 1891. "Poezdka k Karagasam v 1890 Godu [Journey to the Karagas in 1890]." [Papers of the Imperial Russian Geographical Society from the Department of Ethnography] 17 (2): 133–230.

Küçüküstel, Selcen. 2018. "Living with Reindeer and Hunting Among Spirits: The Dukha of South Siberia." PhD diss., Humboldt University, Berlin.

Krivonogov, Viktor P. 2008. *Tofalary: Tri Shaga v Budushchee* [Tofalars: Three Steps into the Future]. Krasnoiarsk: Krasnoiarsk State Pedagogical University.

Mel'nikova, Larisa V. 1994. *Tofy* [Tofas]. Irkutsk: Vostochno-Sibirskoe Knizhnoe Izdatel'stvo.

Noll, Richard, and Kun Shi. 2009. "The Last Shaman of the Oroqen People of Northeast China." *Shaman* 17, no. 1–2: 117–140.

Pavlinskaia, Larisa. 2002. *Kochevniki Golubykh Gor* [Nomads of the Blue Mountains]. Saint Petersburg: Evropeiskii Dom.

Petri, Bernhard. 1927. *Olenevodstvo u Karagas* [Reindeer Herding of the Karagasses]. Irkutsk: Izvestiia Biologo-Geograficheskogo Nauchno-Issledovatel'skogo Instituta pri Gosudarstvennom Irkutskom Universitete [Communications of the Bio-Geographical Scientific Research Institute of the State University of Irkutsk].

Pomishin, Semen B. 1971. "O Transportnom Ispol'zovanii Olenia Tofalarami [About the Transport Use of Reindeer by Tofalars]. *Sovietskaia Etnografia* [Soviet Ethnography], no. 5: 128–131.

Popov, I. I. 1879. "Zapiska o Karagassakh [A note on the Karagas]." In *O Pervykh Priemakh Fizicheskago Vospitaniia Detei u Razdnykh Narodnostei, Naseliaiushikh Irkutskuiu Guberniu* [Methods of Early Childhood Physical Education among Diverse Nationalities Residing within Irkutsk Province]. Irkutsk: Irkutskii Gubernskii Statisticheskii Komitet [Statistical Committee of the Irkutsk Governorate].

Radlov, Vasilii V. 1907. *Proben der Volksliteratur der Türkischen Stämme: Mundarten der Urianchaier (Sojonen), Abakan-Tataren und Karagassen.* Saint Petersburg.

Rassadin, Igor V. 2000. "Khoziaistvo, Byt i Kul'tura Tofalarov [Household, Subsistence and Culture of Tofalars]." PhD diss., Institute of Mongolian, Buddhist and Tibetan Studies, Ulan-Ude.

———. 2005. *Khoziastvo, Byt i Kul'tura Tofalarov* [Tofalar Subsistence and Culture]. Ulan-Ude: Izd. Buriatskogo Nauchnogo Tsentra SO RAN [Publisher of the Buriat Scholarly Center SO RAN].

Rassadin, Valentin Ivanovich, and Svetlana Menkenovna Trofimova. 2018. "Obshemongol'skaia Leksika po Razdelu 'Viuchnyi i Verkhovoi Transport' v Khalkha Mongol'skom, Buriatskom i Kalmytskom Iazykakh [Pan-Mongolian Terminology in the Category of 'Pack and Riding Transportation' in Kalkha Mongolian, Buriat, and Kalmyk Languages]." *Izvestia Volgoradskogo Gosudarstvennogo*

Pedagogicheskogo Universiteta [Communications of the State Pedagogical University of Volgograd] 5, no. 128: 160–163.

Rassadin, Valentin Ivanovich. 1971. *Fonetika i Leksika Tofalarskogo Iazyka*. Ulan-Ude: Buriatskoe Knizhnoe Izdatel'stvo.

———. 1996. *Legendy, Skazki i Pesni Sedogo Saiana: Tofalarskii Fol'klor*. [Legends, Tales and Songs of the Gray Saians: Tofalar Folklore]. Irkutsk: Izdatel'stvo Kommercheskogo Tsentra 'Zhurnalist' [Publisher of the Commercial Center 'Journalist'].

Sharastepanov, D. Ts.-B. 2008. *Oka: Gody i Liudi* [Oka: Years and People]. Ulan-Ude: Respublikanskaia Tipografia.

Shirokogorova, Elizaveta Nikolaevna. 1919. "Severo-Zapadnaia Mandzhuriia (Geograficheskii Ocherk po Dannym Marshrutnykh Nabliudenii) [Northwestern Manchuria: A Geographic Study Based on Observations Made en Route]," *Uchenye Zapiski Istoriko-Filologicheskogo Fakulteta vo Vladivostoke* [Scientific Proceedings of the Historical-Philological Faculty at Vladivostok] edited by S.M. Shirokogoroff, no. 1: 109–46.

Shtubendorf, Iu. 1858. O Karagasakh. *Etnograficheskii Sbornik*. Pp. 1-18.

Stoianov, A. 1980. *Tofalarskie Narodnye Pesni* [Tofalar Folk Songs]. Irkutsk: Irkutskii Oblastnoi Tsentr Narodnogo Tvorchestva i Dosuga [Irkutsk Regional Center of National Art and Leisure].

Tkačenko, I. D. 2010. "Riding Horse Tack among the Cattle-Breeders of Central Asia and Southern Siberia in the First and Second Millennia CE." *Étude Mongolei et Sibériennes, Centrasiatique et Tibétaines*, no. 41: 1–25.

Vainshtein, Sevyan I. 1957. "Ocherk Etnogeneza Tuvintsev [An Overview of Tuvinian Ethnogenesis]." *UZ TNIIYaLI*, no. 5.

———. 1969. *Lichnoe Imia, Terminy Rodstva i Prozvishcha u Tuvintsev* [The Given Name, Kinship Terminology, and Nicknames of Tuvinians]. Moscow: Onomastica.

———. 1974. *Istoria Narodnogo Iskustva Tuvy* [A History of National Art of Tuva]. Moscow: Nauka.

———. 1978. "Etnorafìcheskie Issledovaniav Gornom Altaie i Tuve[Ethnographic Research in the Mountainous Altai and Tuva]." *Polevye Issledovania Instituta Etnographii*. Pp. 90–100.

———. 1980. *Nomads of South Siberia*. Edited by C. Humphrey. Cambridge: Cambridge University Press.

Vasilev, Viktor N. 1910. "Kratkii Otsherk Byta Karagasov [A Brief Overview of Karagas Subsistence]." *Etnograficheskoe Obozrenie* [Ethnographic Review], 1–2, no. 84–85: 46–76.

Wheeler, W. Alan. 2001. "Inalienable Possessions, Identity and Reinventing Reindeer Herding in Northern Mongolia." Paper presented at the 100th Annual Meetings of the American Anthropological Association, Washington, DC, November 28.

Chapter 6

Taking Stock

Tofa Reindeer Herding Today

Alex Oehler

Following the previous two chapters' discussion of Tofa multispecies households, and of horses more specifically, this chapter delves into the ongoing transformation of Tofa reindeer breeding. The information presented here is based on a combination of ethnographic field notes taken during a 2014 visit to Alygdzher's autumn reindeer camp on Dugol'ma Creek, open ended interviews conducted with active reindeer herders in 2014 and in 2018, and key published sources (e.g., Rassadin 2005; Donahoe 2004; Mel'nikova 1994). Most insightful to this chapter are the contributions made by reindeer herder Valentin Tulymov (b. 1984),[1] with whom I met at Alygdzher in June 2018 to discuss the state of Tofa reindeer herding. Tulymov provides the perspective of someone who possesses a deep commitment to the ancestral practice, but who represents a younger generation of herders who began their activities only a decade ago, and who received their training from a generation of herders who were active under the Soviet collective farming system.

The goal of this chapter is to reflect on current Tofa herding practices in light of several historical developments that have co-shaped many Tofas' relations with reindeer. It opens up with a brief summary of the Tofa transition from a nomadic way of life to Soviet sedentarization and its concomitant attempt (and failure) at diversifying the productivity of mountainous taiga households. The chapter then moves on to contemporary Tofa reflections on reindeer domestication as a matter of curiosity: curious as a topic, and with curiosity as a tragic outcome. Caught within the dilemma of curiosity, I examine some of the qualities Tofa herders attend to in their making of a "good home" for domestic reindeer. Specifically, I am interested in the design of calving sites in alpine taiga settings, which grows from a fusion of ancestral and Soviet methods. Reflecting on some of the Soviet heritage of reindeer management in Tofalaria, the chapter reiterates Tulymov's concern about the change in how members of his community relate to reindeer, before concluding with the current Tofa herding cycle.

FROM SEDENTARIZATION TO DIVERSIFICATION

For most Tofas, the transition from a highly mobile lifestyle to a life settled in villages occurred rather rapidly. Over the course of only three years—from 1927 to 1930—the majority of Tofa families exchanged their conical tents for permanent buildings as their primary residences (Mel'nikova 1994, 222). A series of articles from the 1930s, published in the journal *Sovietskii Sever* [The Soviet North], describe some of the nuances of this transition (ibid.). *Integralsoyuz*[2] had constructed subsidized permanent homes at several locations, which today constitute Tofalaria's three Tofa settlements of Alygdzher, Verkhniaia Gutara, and Nerkha. To sustain its new residents, the organization had to import food over great distances and at great cost. Deprived of access to their own animals, and immobilized by sedentarization and full time employment by the emerging collective farm, people were no longer in a position to subsist on their own. The rationale had been that full Socialist collectivization, and economic optimization of taiga resources by way of collective farms [Rus. *kolkhoz*], was possible only after the population had fully settled.

Once families had settled into their permanent homes, they became liable to paying for their imported rations. All food imports arrived in winter on sled, and residents paid for their lot with the proceeds they received from the farm-organized winter fur harvest. However, as the winter hunt fluctuated in its success, communities began to incur increasing debt with the state (Mel'nikova 1994, 223). To help cover the financial deficit, the newly formed kolkhozes began charging hunters for the rental of their transportation reindeer, which led to severe resistance among locals (ibid., 226). The rental fee, however, stood in line with the administration's general aim to diversify the Tofa economy far beyond hunting. In fact, the kolkhoz model sought to maximize the productive capacity of Tofalaria's mountainous taiga zone by collectivizing all clan-based households.

By 1931, each of the new Tofa settlements had their own kolkhoz. Each kolkhoz was responsible not only to meet externally set quotas for fur hunting and reindeer breeding. Quotas for breeding cattle stock and horses were also introduced, and large scale gardening, wild foods gathering, riverine fisheries, and pine nut collection became mandatory (Mel'nikova 1994, 238). Cattle breeding required pasture, of which there was very little in the taiga, gardening was difficult given the high elevation, and fisheries were a foreign concept at the industrial scale. Throughout its existence in Tofalaria, the kolkhoz model led to chronic underproduction, keeping the population in permanent debt to the state. Even the sale of reindeer meat, which began in the 1950s, was unable to alleviate the situation (ibid., 255). In a final attempt to reduce the growing debt, all three kolkhozes were merged into a single *koopzverpromkhoz*[3] in 1967, narrowing the communities' focus to the more feasible activities of fur hunting, reindeer breeding, and nut collection (ibid., 237, 256).

Although the production mandates of the kolkhoz system had been lifted with the transition to a narrower koopzverpromkhoz model, several of its innovations—such as the prolific gardens and dairy cows—were selectively held on to by individual households. Following the collapse of the Soviet Union and its consequent dissolution of subsidies, the residents of Alygdzher were wise enough to major on all possible techniques that would preserve life in an isolated community. At the time of my visits, many of Alygdzher's households relied on their cellars year-round. These were stocked with canned produce from their own vegetable gardens, with dairy products from one or more hay fed dairy cows, with locally caught fish, and with wild berries, onions, and mushrooms. While gardens and limited dairy cattle required a household's year-round presence in the village, many men also kept a firm foot in the taiga.

Tulymov was one of these men. He had recently built a new home for his young family, taking advantage of a government subsidy available to residents of Alygdzher to improve their standard of living. While his wife was cultivating a large garden behind their new home, Tulymov had taken time off from his responsibilities as one of the community's few paid reindeer herders to complete the construction of the new home. He had begun working as an aid to the reindeer herders prior to his military service. Upon returning from the army, he decided to work as a full-time herder. That had been eight years ago. Although he had enjoyed the recent break from herding while working on his house, the reindeer had been on his mind the whole time.

On the Curiosity of Domestication

For Tulymov, as for many other Tofa herders, the word "reindeer" evoked a sense of curiosity. Perhaps this was true in more than one sense. People were generally curious about, or at least took interest in, why reindeer behaved the way they do. However, even more so, the animal itself was famous for its own highly developed sense of curiosity. The desire to investigate, even at the cost of personal peril, was a characteristic that for many herders embodied one of the pitfalls of domestication. Wolves, for instance, catch reindeer by the snout or neck. That is, they attack reindeer head-on. Herders argued that reindeer failed to run out of curiosity. When they did run, they kept turning around to see who was pursuing them. It is in this instance that wolves will charge and attack them. A similar scenario took place when people drove reindeer from one place to another. Jumping up, the deer would repeatedly halt to turn around and investigate their pursuers. This could be problematic, especially when two or more reindeer caravans traveled one behind another. A caravan running in front could easily be startled, causing entanglement with the one following behind.

By contrast, non-domestic reindeer did not seem to stop and turn around when chased. Instead, they sought shelter on cliffs, or they would get away by entering deep snowfields. Although wild reindeer may find themselves cornered on cliffs, in much the same way as South Siberian hunters use exposed cliffs to corner ungulate game, this type of refuge enables reindeer to shake off predators. Deep snow cover dissuades wolves from following reindeer, as reindeer move much faster under such conditions. In most cases, according to the herders, domestic reindeer resorted to neither of these tactics when chased. The emergence of curiosity in domestic reindeer, as opposed to their wild relatives, is also evident in some of the historical records (and archaeological remains) of summer pastures (Mel'nikova 1994, 67–68). Here Tofa herders erected their conical lodges inside corrals to prevent curious reindeer from rummaging through human belongings.[4]

Although there is no record of Tofas ever using their dogs to herd reindeer, trained transport reindeer were quite familiar with Tofa hunting dogs. Upon their return from the autumn reindeer camp to the village, hunters packed their saddled reindeer with winter supplies. At this point, the hunters tied their dogs to the deer, extending their multispecies caravan. If threatened by a dog, a reindeer could easily trample the dog, sometimes even to death. In any case, the reindeer were unlikely to attempt to flee the presence of the dogs. In one instance, where the paths of hunting dogs had crossed with that of a reproductive herd, Tulymov had seen how two herders had made use of their deer's confidence and determination: a cow had been resisting their attempts to tie it down. In response, the men led a leashed hunting dog near the cow's calf. Its mother immediately came dashing from the bushes, ready to attack the dog with its hooves. Although the aggressive defensive response of domestic reindeer may be useful in some instances, herders like Tulymov generally thought of this behavior as a negative outcome of the reindeer-human domestication relationship.

A further trait unique to domestic herds was their sleepiness. Especially older animals exhibit drowsiness, taking frequent naps. After feeding, many of them will bed down in the soft moss of alpine wooded slopes or in the shade of tall trees. Here they snooze, which often turns into a deep sleep. A herder may walk past a napping deer, tapping its antlers or its side. Rather than jumping to its senses, older cows will merely wiggle an ear, without opening an eye to assess the situation. From the standpoint of the herders, such behavior was potentially dangerous. Dull confidence and lack of attention to perturbations, coupled with the heightened sense of curiosity and defensive aggression of the younger animals, served as examples of some of the more counterproductive outcomes of the domestication relationship. Tofa hunter-herders saw none of these traits in wild reindeer.

What made the herders' lives even more difficult was the ease with which hard-earned trust and familiarity could be broken in some of the animals. Locals referred to such reindeer as the "wild ones" [Rus. *dikari* or *dikar'shary*]. A wild one—a *dikar* (singular)—after being startled once, would act out wildly in spite of its upbringing near human encampments and its training with salt from an early age. Once such a deer had been startled, it was nearly impossible to manage it as part of a herd. The dikar would resist with its hooves all human attempts of approaching. At the time of my field visits, local herders were reporting the number of dikari as especially high.[5]

COMING INTO THE WORLD: ON DESIGNING CALVING SITES

Beginning in April, herders rounded up the reproductive herd [Rus. *matochnik*] consisting of reindeer cows and their young offspring. Meanwhile, they also prepared the poles [Rus. *solbaki*] to which the cows were going to be tethered (cf. Mel'nikova 1994, 68). Solbaki are the trunks of felled trees, most commonly larches. Herders stripped the trunks of their branches and crowns, and after debarking them, drilled holes through the stems in multiple locations to accommodate tethering ropes. Then they evenly distributed the poles in the pasture, allowing for equidistant tethering of animals with non-overlapping grazing radii. A *solbak* (singular) measures 5 to 6 meters in length, accommodating three to four reindeer. To prevent the animals from dragging it around, the poles were held down by sharpened pegs, hammered into the ground at an angle. By removing the pegs, the herders could easily reposition the poles by a few meters each day. All the cows expected to give birth that season were tethered here, and once they had calved, their young were tethered to ensure the mothers stayed at camp. Before any of these activities could commence, however, the herders had to identify a suitable calving ground.

It was not a simple matter to find the right kind of calving ground [Rus. *mesto provedeniia otela*]. From approximately 2012 to 2017, the calving ground for the Alygdzher herd had been located on Dugolma Creek, just below the autumn roundup camp—a two-day horse ride from Alygdzher. Over the span of these six years, many calves had been born on the Dugolma. However, herders generally preferred not to use a calving site for as many consecutive years. Given the slow recovery of lichen, extended re-use can lead to feed shortages at a time when consistent nourishment is of utmost importance (Herder, Kytöviita and Pekka Niemelä 2003). At the same time, it is exceedingly difficult, with a small group of people, to drive a herd of 100 head to a new location. Such an endeavor requires a sizeable team of workers, and springtime was one of the busiest periods for people at Alygdzher. It required at least four weeks to successfully redirect the reproductive herd. During this time, the herders had to locate the animals in the forest, rounding them up multiple times. In recent years, no corrals had been in use for reindeer, further complicating the roundup. After the animals were gathered, the herders drove the entire herd to the new location, ensuring no stragglers were left behind.

According to Tulymov, in an ideal calving ground, the snow is less than waist deep upon arrival. However, even under heavy snow conditions, what really matters is reliable feed. If a good amount of feed is present under the snow, the herd may calve even in deep cover. The cows will bury themselves deep into the snow. Over time, they knock off snow layers to access the lichen bed on which they are standing. Preferably, this consists of a lichen carpet of now less than 30 centimeters in thickness. Every morning the herders will move the cows' poles by a distance of some two meters, allowing them to access fresh feed for the day. In the evening, the herders move the poles again by another 2 meters for nighttime feeding. This ensures that the lactating cows are always well satiated, able to sustain themselves while producing sufficient milk for their young. A succession of quick freeze-thaw-freeze periods in early spring can result in the top layer of snow to form an impenetrable ice crust.[6] Where this happens, it is likely that a larger number of animals will die of starvation.

When selecting a new calving ground, herders must look for a nearby water source, as cows and calves require regular watering while being tethered. The site should also be void of deeper ruts or naturally occurring trenches. If it so happens that after the herders' roundup some cows remain unaccounted for, it is likely that they will muffle their young inside these cavities in the ground. It is a strenuous and time-consuming task for herders to locate and extract these calves. Generally, a calving site should be located on dry ground. As the heat of day gradually increases later in the spring, ideally selected sites will retain intermittent snow patches as a source of easily accessible hydration. Snow patches also help animals to cool off near the end of the calving period in June. In 2018, sixty cows had already given birth by the middle of June, and they had all been able to secure their own water, speaking for the quality of their calving site. Finally, a calving site must have good visibility. Thick forest provides shelter to predators, and wolves and bears had been a recurring issue for herders in Tofalaria.

Over the late spring period, all the yearlings [Rus. *godovalye*] roam freely within the general vicinity of the calving grounds. Periodically, herders will round them up for tethering, to ensure they remain accustomed to the reindeer bridle [Tof. *mungui*] (cf. Mel'nikova 1994, 71–72) to which they were first introduced in the previous year at the same location. The bridle consists of a wooden arch with a rotating wooden peg to which the tether attaches (Rassadin 2005, 32; Mel'nikova 1994, 70). The arch goes under the lower jaw of the calf, where the bridle holds it in place. The wooden peg protrudes slightly from the inner side of the wooden arch, pressing lightly on the calf's tongue muscle from the underside of its lower jaw. The protrusion slows its suckling, which in the past secured a greater milk yield in the dairying process. Even when the cows are not milked, the simple mechanism of the rotating peg prevents the lively calf from becoming entangled in a twisted tether. All newborn calves are tethered for the first time at the age of four days.

The fourth day marks the start of a calf's training, which at first consists of catching it in the morning to be tethered for the day, and releasing it in the evening. Conversely, the mother is released from her tether in the morning. She is then allowed to graze freely throughout the day, always staying near her young. Come evening, she is tethered again, with her young staying nearby. For the first few days, many of the calves attempt to free themselves from their tethers, wildly pulling on their mungui and tether. During this time, the herders continue to select new grazing spots for their rotating mother-calf couples. In the past, the mungui was essential to the milking process. A calf, tethered near its mother, would stimulate her udder by prodding it repeatedly. After a few draws, the calf was removed and milking commenced. During milking, the cow stood tethered to a *mangush*, which was a log with one of its ends buried in the ground and the other raised above ground by one meter, held up by two sticks (Mel'nikova 1994, 70). The structure formed a kind of triangular saw-horse-like tethering pole, ideal for milking. None of these structures were in use during my fieldwork, as reindeer dairying had long been abandoned.

Due the very slow growth rate of reindeer lichens (e.g., Herder, Kytöviita and Niemelä 2003), it was difficult to secure fresh patches in a site that had been in use over several years consecutively. To secure thick carpets, ideally a site lies dormant for several decades before it is reused. Seasoned Tofa herders had memory of the migration routes that kolkhoz reindeer brigades had chosen during the Soviet period, maximizing the carrying capacity of the taiga between Alygdzher and the borders to Buriatia and Tyva. Remnants of horizontal tethering poles, as well as the remains of former corrals, were testimony to lichen pasture rotations some of which, according to the herders, went back as far as the interwar years (1922–1941).[7] This period represented the early days of Soviet collectivization, a time when not all Tofa herding families had settled into villages yet. Although present day herders were very much rooted at Alygdzher, their search for suitable pastures continued to rely on a knowledge of where their ancestors had grazed the deer, even as far back as pre-settlement times. In fact, it was clear to everyone that the territories most commonly frequented prior to Soviet sedentarization were now among the lushest.

According to ethnohistorian, L. V. Mel'nikova, the integration of reindeer into the kolkhoz system had resulted in people being ever less willing to travel to distant pastures (1994, 258–259). Based at Alygdzher, and with relatives tasked with activities in the village, there existed a clear incentive to reduce the commute between herd and village. Inevitably, this had devastating consequences for the reindeer, as pastures were being selected for their proximity to settlements, rather than for their richness in lichen. In the end this development also meant that all the lichen pastures that did surround the villages, including Alygdzher, became exhausted. The consequences of this tendency were still being felt during my stay.

Compromising between distance and fodder yield, in the spring of 2018 the herders moved to a new calving ground on Khailama River. The majority of the animals had stopped moving in the direction of the old site, confirming its exhaustion. The new site was located approximately 10 kilometers from the old one, a distance small enough to cause some animals to stray back to the previous site. This straying, which is likely based on migratory habit or memory, added to the herders' workload as they strained to prevent such regressive herd movements, regularly driving animals back to the new site. By insisting on the new location, the herders were helping establish memory of a new migratory pattern in their herd. The new area was actually quite old, and people knew it as Trekhstrelka, located on the Koktei vertex near Ozhigaii and Bolbyrbianykh streams at an elevation of 1,810 to 2,258 meters.

Here in the higher mountains the weather changed quickly, but Tofa herders generally looked forward to poor weather conditions in spring. Reindeer were thought to calve best under precipitation, whether snow or rain.[8] Some herders explained this as a natural phenomenon that helps prevent predators from detecting the location of the calving animals. The precipitation helps wash off the blood lost during the birth, naturally concealing the cow-calf couple, thus reducing their risk of predation. Local hunters observed a similar phenomenon in wild red deer and Manchurian wapiti, both of which often have their young under conditions of precipitation. The herders reasoned that even if their reindeer were domesticated, the urge to hide their offspring remained strong, and calving under precipitation was a feature their reindeer had retained.

THE HERITAGE OF SOVIET MANAGEMENT

Unlike the present day, during the Soviet era reproductive and transportation herds had been kept strictly separate. The castrated and trained riding deer had been herded closer to the settlement for ease of access, while the larger reproductive herd with its cows and yearlings was kept further away to benefit from richer and more flexible feeding arrangements. Tulymov's father had worked in the "transportation herd" (Rus. *transportnoe stado*) for many years. The herd had received its name from the fact that it was employed by geological expeditions to transport people and equipment into the mountains.[9] The transportation herd was also the primary means of all kolkhoz-employed Tofa hunters to reach their assigned winter hunting parcels. It had been stationed on Utiug Mountain at 1,780 meters elevation, just south of the village. From here, it was easy to equip expeditions departing from Alygdzher. Supplies that had reached Alygdzher on sleds over the frozen river in winter, or which had been flown in during the summer, were packed on the reindeer prior to departure into the taiga.[10] The reproductive herd had stayed further afield. At some point in time it had been stationed near Peak Asardik. Workers from both herds had been on constant rotation, and at its peak, some forty people were employed as herders and helpers, according to Tulymov.

Each herder had been responsible for a group of reindeer, which he would pass on to the next herder at the end of his shift. A typical group consisted of twenty to thirty animals, all of which were assigned a common color. A thread of that color was then tied to the reigns of each animal, preventing confusion with other members of the larger herd. At the beginning of summer, a herder who had taken on a group from the previous shift in early spring would pass on the same number of animals, including all their newborn. The summer herder then passed the group back to his colleague in autumn. Along with his group of deer, herders passed back and forth a logbook in which they tracked each animal by its assigned number. Each animal's number was shorn into its coat on both sides, making it easily countable from a distance (see Figure 6.1). In addition, each reindeer carried an earmark. Earmark records were also being kept in the logbook. At the end of a work shift, the incoming herder would take note of all recorded changes in the logbook. The logbook was then passed on to the senior herder who subtracted from the herders' salary in accordance to the number of animals lost during each shift.

Tulymov recalled his father's stories of the state of the herd in the late 1970s, when the wolf population had become a particular challenge. In response to severe wolf attacks, which had shrunk the reproductive herd to some 200 animals, the herders had driven the reindeer toward Alygdzher. Their aim had been to shake off the wolves, and they had largely succeeded

Figure 6.1 Oehler with Reindeer No. 9, Tofalaria. *Source*: 2014, K. B. Klokov.

by the time the herd was nearing the village. Yet, a number of deer had fallen to the wolves during the trek, in spite of the herders' use of wolf poison. By the mid-1980s, the reproductive herd had recovered, sheltered as it had been by proximity to the village. By now, it was counting some 600 head. Having regained size, the herders proceeded to drive the animals to a higher elevation on Khailama River, where in subsequent years the herd grew to around 1,000 head. The herd's large size made it feasible to export some of the live animals and to slaughter others for meat sales, as had been the practice earlier on in the 1950s.

By the late 1980s, many other reindeer breeding regions throughout Russia were interested in strengthening and diversifying the genetic pool of their reproductive herds by introducing tall Karagas bulls from Tofalaria. Tulymov recalled bulls being shipped to Chita Oblast', Khabarovsk Krai, and as far as Canada at the time of his childhood. Groups of 100 and 200 reindeer had been sold at a time. Loaded onto helicopters, the deer were flown to their destinations. During these years of the late Soviet period, the Alygdzher herd had peaked at around 1,500 animals, making it feasible to slaughter between fifty and sixty animals for meat each autumn. Herders castrated and trained for riding the stronger three-year-old males. More meagre looking two and two-and-a-half-year-old males were also castrated to prevent them from affecting the herd's gene pool. Finally, they would slaughter all the limping or askew looking animals for meat sales.

A CHANGE IN RELATIONSHIP

From the perspective of several contemporary Tofa herders and hunters, including Tulymov, the collectivized method of reindeer breeding during the earlier Soviet period represented an era during which people had been able to retain at least some of the pre-sedentary relationship with their deer. Not only were many young Tofa men able to make a living from their herding responsibilities as part of the collective farm, but by so doing, they had been able to spend time near their elders. Many of these elders had grown up in remote taiga settings. Some of them possessed memory of a time preceding the dispossession of their hereditary clans' reindeer. With summers regularly spent at reindeer camp, children and teenagers had remained exposed to their elders' ways of relating to reindeer, even if during the school year their primary homes were now in the village.

Contemporary Tofa herders, several of whom were born between the early to mid-1970s and 1980s, enjoyed recalling their grandparents' and great grandparents' stories. These were accounts from the cusp of collectivization, a time when Tofas still had a strong sense of belonging to distinct clans [Rus. *rody*], each of which would once have held between thirty and forty, or as few as twenty or twenty five, reindeer. Such small numbers had enabled people to remain highly mobile. Contemporary herders, like Tulymov, felt rather envious thinking about the ease with which a herd of this size would be rounded up. However, not only in terms of practicality had things changed with collectivization. Smaller numbers had allowed people to be more intimate with their animals. Both men and women, including children, had engaged with their herds in a hands-on way, included milking.[11] In the minds of today's herders, this intimate proximity to a small group of animals had produced a radically different kind of relationship between people and animals than what active herders were witnessing among the youth of Alygdzher in the post-Soviet present.

The former bond between Tofas and their reindeer had also been an outcome of extensive shared hunting experiences, which had persisted into the Soviet era. In the sovkhoz, according to Tulymov, individual herders were allowed a month of subsistence hunting each year. For this purpose, they each rented a group of castrates from the transportation herd. Immediately upon their return, however, the men were sent back to resume their posts as herd workers. While their employment secured them a salary, this new professionalization of an old practice had driven a wedge between the previously inseparable practices of hunting and herding. The very purpose of reindeer in Tofa society had been to move the household through difficult to access terrain where families would set up temporary camp for subsistence hunting. Reindeer had carried infants in specially mounted cradles, and children grew up playing alongside their families' reindeer. Under such conditions, it was a given that all household members would be deeply familiar with the individual personalities of each of their deer.

Such intimacy was lost with the collectivization and amalgamation of family herds, and with the rigid segmentation and professionalization of hunting and herding tasks.[12] Not only had the herders' family life become increasingly detached from the life of the herd, but also the sheer size of the herd had made it impossible to be intimately familiar with each animal. Furthermore, the herding process no longer connected to other household activities. Instead, it had become a salaried routine in which one's proficiency was metered against the satisfaction of one's superior rather than the expediency of one's joint household. Hunting, herding, and other chores of the household had become disjointed, and the intuition and spontaneity of the old herding style had been lost to a highly structured, often impersonal, management style that is well summed up in Tulymov's recollection of his father's roundup activities in the late 1980s:

> Having tied together two or three cows, the yelling herder would drive the trio to camp [before heading out again to repeat the process]. As one of the animals would pull, the other two would be chasing after it. With more and more animals arriving at the camp, the first, second, and third groups were driven into the corral, where the herders quickly counted them. Then one group would be released, and by morning another would be released. A week later [all the animals] would be rounded up again. If the senior herder demanded another roundup for the next day, then the whole procedure would be repeated again the next day.

Since the demise of the Soviet Union, and the concomitant collapse of the kolkhoz and zverpromkhoz systems, the herd had experienced several waves of growth and decline, and with it had come and gone several collective management models. By 2018, however, there existed no reliable funding structure to support committed full time herders. More importantly, there seemed to be no desire among the local youths to spend time herding reindeer in the taiga. For Tulymov, lack of pay was not the only reason for the disinterest.

> Even in those [Soviet] days, there had still been around many elders. Tofa elders of the past were used to this [remote] life. They had lived out there, training the teenagers. Consequently, their children, grandchildren, and great grandchildren had all grown up out there with the herd. That is how they had been moving.

> The elders died off, the young people all transitioned to a sedentary life, the koopzverpromkhoz fell apart, consequently that power, and those times went away and all that sank into oblivion. We have no way of dealing with this. How would one do it?

Not only the monotony of the Soviet style of herding, but also the rough living conditions and long absences from the settlement, were unpalatable to young people now. A month-to-month rotation system had recently been proposed, but there had been no takers. To Tulymov's disappointment, the younger generation was far more interested in electronic gadgets and the Internet. However, for Tulymov, the root of the problem lay not in lack of pay or in the electronic preoccupations of a younger generation. At the heart of the issue lay a change in the relationship with the animals. What people had lost was a deep longing to be with the deer.

For Tulymov, the relationship with reindeer had been broken not by Soviet collectivization. Sure, sedentarization, collectivization, and professionalization had all altered the rhythm of people and animals, and it had affected who was working with whom. But it had not entirely wiped out the desire, or habit, of people to spend time in the taiga. The connection to their elders had been preserved, and in the isolation of the taiga, elders had been able to speak their ancestral language. The most cardinal change in the human-animal relation, according to Tulymov, had been a much more recent phenomenon of the 1990s. Referring to his generation, Tulymov said:

> We are among the last children of Tofalaria. The language went away. All we have left are the reindeer. We have nothing else left in Tofalaria. This is very regrettable. To this day, it torments me. [. . .] If [only] all things had not fallen apart . . . That is what it all hinged on: the nineties—when in the 1990s everything fell apart. Had that not happened . . . if that generation had not been let go of . . . Even in relation to our Tofa language . . . This was a big loss.

Tulymov saw a direct connection between the Soviet work ethic (i.e., the state's insistence that all people show up at work regularly and on time) and a person's continued relationship with the taiga and its animals. Once the framework of the zverpromkhoz had fallen away, there was no direct insistence on people to spend extended periods in the taiga. Although the Soviet education system had not served well the needs of the ancestral Tofa language, the workings of the zverpromkhoz had indirectly secured some preservation of the language through the contact it had facilitated between younger herders and their elders. By the 1990s all this had disappeared. Without subsidies or a market for reindeer products, and with little hope for a future in reindeer, the next generation of would-be herders had missed their training and lost their language. Yet, the reindeer were still there.

THE PRESENT CYCLE

Following the demise of the subsidized state run system of reindeer herding, residents of Alygdzher became increasingly dependent on subsistence and trophy hunting.[13] In 2018 all households relied on a steady meat supply from the taiga, as the import of foods was still prohibitively expensive. Most households had one or more dedicated hunters, while others traded their labor for meat. In 2014, approximately 100 residents were actively involved in annual winter hunts. Correspondingly, in 2018, the reindeer herd consisted of approximately 100 castrated and trained transportation deer, along with another 100 or more animals in the reproductive herd.[14] Many of the trained riding males [Tof. *chary*] bore the initials of their operators. However, only hunters who maintained access to hereditary hunting grounds at higher elevations and in steeper terrain relied on reindeer. All other hunters with access to less difficult territories were increasingly using snowmobiles and other motorized means for transportation.

As the pay for dedicated herders had disappeared, and with ever fewer capable and willing herders at hand, the shrinking reproductive and transportation herds had been amalgamated into one. Having all the animals combined into a single herd made it possible to oversee and co-manage their annual migration cycle with a small number of dedicated herders. At the time of my field visits, hunters from the village had set up a rotational herding roster, according to which they staggered their reindeer camp visits to pick up their riding deer in autumn. As hunters arrived at camp, they took up herding responsibilities, allowing the full time herders to leave camp to follow other tasks.[15] As hunters arrived at camp, they would sign out their deer, keeping record in a hunter's logbook left inside the camp cabin.

Following the winter hunt, and after the return of the transportation deer to the reproductive herd in the forested hills in March, the joint herd would be gathered and moved up to the calving grounds. If all went according to plan, the herd was met here by volunteer calvers [Rus. *teliatniki*]; young men selected by the head herder. The calvers would then help the herders with the tethering rotation. After the calves were born, and after their mothers had regained some strength, the calvers accompanied the herd on its way to higher elevation. This movement to the summer pastures was not rushed. As Tulymov explained: "By mid-July they go up the Oab River. We release the cows along with their young, and every second day we go. We can travel for a day, two, or even three. Then they arrive at the encampment." By driving the herd only every second day, the cows and their fragile young were given plenty time to feed and rest. Depending on the chosen route, the transition from calving grounds to summer pasture could take up to a week.

As the calvers prepared to move uphill from the calving grounds, the herders would descend to the village, where they loaded up on summer camp supplies, including salt. Dropping off their supplies at the summer pasture, they also set up camp for the season. At this point the calvers and the reproductive herd would already be on their way up the mountain. As a herder waiting for his calvers, Tulymov explained: "We observe as they approach the camp. As they arrive, everything has been prepared for them. As they come, they know that we are here. Later we keep checking on [the herd], as we now live at this camp. We check on the herd, but we also monitor for predators. You leave camp in the morning and spend the whole day out there looking after the herd at the top of the mountain." Herders were not only waiting for the reproductive herd to arrive. They were also keeping an eye out for the transportation deer that would make their own way to the summer pasture, after they had been released near the village in early spring. Throughout the summer, the campsite was staffed by two or three people, be they herders or helpers.

In autumn, the herd gradually began its descent, in search of mushrooms. With them, the herders packed up camp and followed the herd. Beginning in August, and into early September, the herders castrated the males. Mosquitoes and flies had died down by this time, reducing the risk of infection. Once the herd arrived at the autumn camp, the females were gathered. The goal would be to keep them near the camp until first snow. Hunters from the village would now arrive to search for their trained riding deer. Because the transportation herd was moving along with the reproductive herd, male castrates could be located quite easily. As the cows were lured with salt to the center of the campsite, hunters would sort out their transportation deer, tying them down just behind camp. When a hunter had gathered all of his castrates, he passed the hunter's log on to the next arrival and descended to the village with his caravan. From the autumn camp, hunters and herders would monitor the herd's movements to determine whether predators were nearby. With the snow cover deepening at higher elevations, the remaining stragglers would finally arrive at camp.

Their wild looking antlers immediately gave away the new arrivals. Antler velvet was shedding by this time of year, hanging down in shreds. As animals approached the salt covered rocks at the center of the campsite, two herders would grab one cow at a time, tying her to a tree to clip her tines. While one herder held the cow in place, the other took a handsaw and selectively shortened each antler, trimming off sharper tines. It was important that some antler remained for snow scraping in winter, and for defense against predators in spring. Herders also removed the sharper tines to prevent injuries from being inflicted on other animals, including herders and hunters. Heavy bull antlers were clipped to ease the animals' movement and to prevent entanglement with other bulls during the rut (see Figure 6.2). Following the procedure, the cows and bulls dispersed into the wooded hillsides in their family groups.

The rut, which coincided with the autumn roundup, played a role in the composition of these family units. With all the females milling about camp, the bulls [Rus. *khory*], who previously had moved in a single group, were also drawn to camp. Here they split into several smaller groups, one of which consisted of the one-year-olds [Rus. *molodniak*]. Mature bulls could be seen visiting camp, on occasion clashing each other's antlers, as they fought over specific females gathering at the salt licks. Once the mating order had been established, females left the camp along with their chosen mates and their young from the previous year. Together these family units gradually spread out into the surrounding taiga, beginning in the upper reaches of smaller tributaries feeding the stream that runs through the autumn camp. This dispersal marked the beginning of winter herding.

Winter herding consisted of a combination of the former Soviet pattern described above and a newer and more invitational style. Tulymov explained: "We already know where [the deer] frequent, along which streams, and so we follow them around. Every now and then we will catch five or six of them and take them down the hill to the campsite." At this point, the herders selected campsites in relation to where the dispersed herd would be moving.

Figure 6.2 Oehler Sampling Blood at Ear Notch, Tofalaria. *Source*: 2014, K. B. Klokov.

In following the herd, the hunters had access to a series of hunting cabins located throughout the area:

> Here we put salt on the rocks. Other deer will smell their tracks and run after them. They will spend a day or two around the campsite. If they have not come, then the herder will set out to find them, scattering some salt near where they are, so they too would eat some salt. He will use this opportunity to check on them and to count them. Although it is difficult to really count them like that. But he will look around and read their tracks.

The skill of reading herd movement and group size from tracks is an example of the frequent combination of hunting and herding expertise in the Saians. Track reading was a common practice in hunting wolves. Hunters would read the tracks of wolf packs frequenting winter pastures. From the tracks they could tell how many wolves were in the pack and of what approximate age they were. This gave the herder an idea of the pack composition that season. The same skill transferred to taiga herding. Even if a herder could not reach a particular group of reindeer, he could count them by their tracks. If he was lucky, he could even identify the family unit. Similar methods would have been used in Soviet years, but the current smaller herd size made this method more effective.

THE RUSSIAN FACE OF FUTURE HERDING

Although the future of reindeer herding may have looked bleak in the 1990s, the following two decades' heavy reliance on subsistence and trophy hunting had played an important role in the survival of Alygdzher's relationship to its herd. Increasingly, however, those descending with caravans from the autumn camp were referring to themselves as Russians or as "*metisy*," rather than as Tofas. *Metis*, in this context, referred to individuals of mixed Indigenous-Russian descent.

With the formation of Tofa villages, Russian families had joined the kolkhozes early on, resulting in countless intermarriages (cf. Krivonogov 2008). By 1950, seven Russian families had settled among thirty-seven Tofa families in Alygdzher, and by 1964 there had been twenty-five Russian families living alongside sixty-two Tofa families (Mel'nikova 1994, 238).[16]

In the post-Soviet context of necessitated subsistence hunting and black market trophy retail for cash flow, reindeer continued to provide some of the most reliable means of transportation for all residents, regardless of ethnic background. It followed, according to Tulymov, that eventually every hunter would have to volunteer some of his time to reindeer herding, irrespective of ethnicity. In 2018 all but one of the hunters volunteering their time at the reindeer camps were of Russian or Métis descent. According to Denis Shulaev, a self-identifying Métis hunter from Alygdzher, "Russian genes overpower Tofa genes every time," meaning that phenotypically every Métis person looked Russian. This Métissage was commonly summed up by the phrase "we have long been Russified" [Rus. *my davno obrussilis*].

Although Tulayev regretted that, "All we [Tofas] have left are the reindeer," for Shulaev the future of reindeer herding in Tofalaria had a Russian face. "They say, 'as long as the reindeer are alive, Tofas are alive.'" "But," Shulaev continued, "I think, pretty soon Russians are going to run the herd." Coming from a tradition that has valued selective breeding for particular attributes in reindeer—such as the famed height of the Karagas breed—herders often drew parallels between the deer and their people. Future ethnographic research will have to examine how Tofa-Russian Métis hunters and herders make use of their involvement with reindeer in defense of their Indigenous status, which in Russia continues to be tied to "traditional" land use.

NOTES

1. This is not his actual name. I use pseudonyms for the herders mentioned in this chapter to protect their identities.

2. This abbreviation translates to "union of integrated hunting and fishing cooperatives" [Rus. *Soyuz integral'nykh promyslovo-okhotnich'ikh rybolovetskikh kooperativov*].

3. *Koopzverpromkhoz* roughly translates to "cooperatively held wild animal products farm."

4. During my autumn reindeer camp visit in 2014, I witnessed the curiosity and fearlessness of reindeer yearlings firsthand, as animals would rummage through my tent and belongings throughout the day, even after repeatedly being driven away.

5. It is difficult to say what had led to this perceived increase in dikari. One possibility is the reduced number of people involved in the herding process. From this perspective, it could be argued the Tofa herd was becoming less domestic.

6. This was a common scenario, historically experienced by other close range reindeer herders, such as the Sámi of Scandinavia who commonly felled trees to access arboreal lichens (e.g., Berg, Gunnarsson and Ostlund. 2011). I did not encounter signs of this systematic felling practice in Tofalaria or Oka during my fieldwork.

7. In Russia, the interwar period generally refers to the years between the end of the Civil War following the October Revolution and the beginning of The Great Fatherland War (referring to the Eastern Front of the Second World War).

8. Changes in the timing and synchrony of the peak calving period have also been found to correlate with climatic changes, particularly a reduction in precipitation in the months preceding the calving period (Paoli et al. 2018).

9. Kolkhozes and state farms of the Altai-Saian region commonly divided their herds into transportation [Rus. *transportnye*] and reproductive [Rus. *matochnye*] herds.

10. Local gold mines were supplied by horses (pers. comm. Tulymov 2018), and reindeer caravan use came to an end when geological exploration began to rely on helicopter services (Mel'nikova 1994).

11. Reindeer dairying had been part of the kolkhoz system of herding (see Fondahl 1989).

12. In 1950, the kolkhoz permitted individual households to operate up to fifteen of their own reindeer for subsistence purposes (Mel'nikova 1994, 241). However, managing a small herd from the village would not have resembled pre-revolutionary mobile Tofa herding.

13. With few employment opportunities in Alygdzher, many hunters relied on hunting Siberian musk deer (*Moschus moschiferus*) for its highly marketable preputial glands (see K. B. Klokov, chapter 5, this volume).

14. Some sixty head were also being kept by a family in the village of Gutara.

15. Given the poor pay for the full time herding position, hunters agreed to contribute some of their own time during camp visits.

16. At Verkhniaia Gutara two-thirds of the residents had been Tofas and one-third Russians, while five Russian families had resided among twenty-nine Tofa families at Nerkha between 1951 and 1958 (Mel'nikova 1994, 238).

REFERENCES

Berg, Anna, Bjorn Gunnarsson, and Lars Ostlund. 2011. "'At this Point, the Lichens in the Trees are Their Only Means of Survival': A History of Tree Cutting for Winter Reindeer Fodder by Sami People in Northern Sweden." *Environment and History* 17 (2): 265–289. doi:10.3197/096734011x12997574043044.

Donahoe, Brian R. 2004. "A Line in the Sayans: History and Divergent Perceptions of Property Among the Tozhu and Tofa of South Siberia." PhD diss., Indiana University.

Fondahl, Gail. 1989. "Reindeer Dairying in the Soviet Union." *Polar Record* 25, no. 155: 285–294. doi:10.1017/S003224740001946X.

Herder, Michael Den, Minna-Maarit Kytöviita, and Pekka Niemelä. 2003. "Growth of Reindeer Lichens and Effects of Reindeer Grazing on Ground Cover Vegetation in a Scots Pine Forest and a Subarctic Heathland in Finnish Lapland." *Ecography* 26 (1): 3–12. doi:10.1034/j.1600-0587.2003.03211.x.

Krivonogov, Viktor P. 2008. *Tofalary: Tri Shaga v Budushchee* [Tofalars: Three Steps into the Future]. Krasnoiarsk: Krasnoiarsk State Pedagogical University.

Mel'nikova, Larisa V. 1994. *Tofy: Istoriko-Etnograficheskii Otcherk* [Tofas: An Ethnohistorical Study]. Irkutsk: Vostochno-Sibirskoe Knizhnoe Izdatel'stvo [East Siberian Book Publisher].

Paoli, Amélie, Robert B. Weladji, Øystein Holand, and Jouko Kumpula. 2018. "Winter and Spring Climatic Conditions Influence Timing and Synchrony of Calving in Reindeer." *PLOS ONE* 13 (4): e0195603. doi:10.1371/journal.pone.0195603.

Rassadin, Igor V. 2005. *Khoziastvo, Byt i Kul'tura Tofalarov* [Tofalar Subsistence and Culture]. Ulan-Ude: Izd. Buriatskogo Nauchnogo Tsentra SO RAN [Publisher of the Buriat Scholarly Center SO RAN].

Part III

DUKHAS, KHÖVSGÖL AIMAG, MONGOLIA

Chapter 7

Householding as a More-Than-Merely Human Improvisation Amidst Manifold Frontiers in a Boundless Home(land)

Nicolas Rasiulis

What does making a home and living on the move with individuals of other species consist of? The household stands out as a salient locus of activity and experience with which such a line of questioning might be effectively "sniffed out" (Ingold 2018, 118), and perhaps even nearly answered. Approaching households in this manner begs further lines of inquiry: Who and what composes households, what is the context of this composition, and what does this entail?

My aim with this chapter is to give a technical account of the ontogenesis of households among Dukhas in a fashion which renders not only the affective, experientially rich living atmosphere of householding in northernmost Mongolia, but also the situational dynamics and compositional tenor permeating and breathing life into households and homeland(s). With this so-called "technical" account, I follow contributors to Frontier Encounters (Billé, Delaplace, and Humphrey 2012) in pursuit of "a technological approach to the study of borders" (Delaplace 2012, 1; original emphasis). By emphasizing "material culture, or technology, that is [. . .] 'the particular domain of human activity immediately aimed at action on matter' (Lemmonier 2010, 684–85)" (Delaplace 2012, 2), we might attend to widely neglected "analytical benefits of considering the border itself as a technique" (Idem, 3; original emphasis).

In this chapter, borders, or perhaps more fittingly frontiers, are thought with not only as spaces of demarcation (i.e., between dominions of political territorialization, and between domains of domesticity, territory, tenure, mastery, etc.). Frontiers are also porous, ambiguous places of relentless perdurance in happenstance with continual change at some of the furthest reaches of state territorialization, of neoliberal market dynamics, of disparate yet increasingly consolidated constellations of multi-stakeholder nature conservation endeavors, and of concurrent commodification.

Many of the frontiers in northernmost Mongolia are at the forefront of change. These frontiers, where particular states of affairs somehow coexist with, subvert, and partially yield to a broad yet distinct change, are themselves characterized by continual change in composition, dynamics, and so on. Although there are changes in local ecosystems brought about by climate change and diverse local resource exploitation, which are intensifying and complexifying in certain areas throughout the area, much literature concerning pastoralism in Central, Inner, and North Asia has posited that change is in the post-socialist present now most rapid and substantive in social domains (Khazanov 1998; Lancaster and Lancaster 1998; Krupnik 1998; Humphrey and Sneath 1999). In the rapidly self-exceeding, increasingly and expansively shared "present moment" that is "increasingly narrated as one of globalization and, with more intensity in the last few years, also as one of crisis" (Blaser 2010, 1), pastoralists' abilities to adjust in relation with both ecological and social systems is as necessary as ever.

Among Dukhas, the most prominently felt and consequential front of change is not only social and ecological, nor is it only particular or only broad. It is the transnational sort of site where, through legislation and consequent logistical implementation, rulers and designers of social-ecological systems partition off areas from one another into units of wilderness using different labels each associated with its own degree of severity of conservation. According to the label, the territory is further partitioned, in this instance from human, dog, and domestic ungulate interactions themselves categorized according to (claimed) severity of ecological impact. Nonetheless, persons, relations, and lifestyles relentlessly perdure beyond the delineations described by such partitions. In attending to Dukha lifestyles and adjustments to these partitions and other major changes I am compelled to approach change itself as manifold, multidirectional, and multiscalar. Change is a nexus of overlapping and entangling domains such as global conservation, particular ecosystems and emplaced spiritual ecologies, some of which might stretch or exceed those of sociality or ecology.

While focusing on the household, this chapter inevitably opens up—just as the household does—to the broader material, social, and spiritual meshworks and processes in which households and householders emerge and carry on together with the boundless taiga community; to the compartmentalized Tengis-Shishged National Park (TSNP) framework of territorial regulation and regulatory implementation; to the frontier of modern state territorialization and neoliberalism.

In the pages that follow, I elucidate some of the concepts and the general theoretical tenor which permeate my technical account of the ongoing emergence of Dukha households, before proceeding with the account itself. This account traces the variegated, situational process of householding in the taiga across disparate yet entangled scales and domains. First, Dukhas' homeland is attended to at the scale of an ecologically and spiritually relational territory. Attention is then turned to the embeddedness of the taiga and its inhabitants in ongoing social, economic, and political processes at broader scales of action, consequence, and analysis. Before concluding, scales of actual households and aĭl groups (i.e., household clusters settling and herding together) come into focus more clearly, regarding both intra and interspecific compositions, endeavors, and adjustments in and out of the taiga. This account is largely informed by my nearly seven months of anthropological fieldwork among Dukhas in both East and West Taigas, and—to a lesser extent—in the nearest ecotone and steppe areas, which I conducted in autumn and winter, 2014, and in summer, 2018.

CONCEPTUAL TIMBRES AND THEORETICAL TENOR

Though the technical account presented in this chapter draws on ideas grown by various people in and out of both the taiga and Inner Asia, I do not attempt to convey either an emic or an etic representation of local systems of meaning, the likes of which might pit self-representations in lockstep with contended discourses. Instead, this chapter is written with a spirit and intent of partially rendering some of my own "sympathetic correspondence" (Ingold 2015, 156; emphasis added) and concurrent "contamination" (Blaser 2010; emphasis added) with the taiga and many of its diverse inhabitants, within and beyond the porous confines of humanity.

To some extent, the concept of "household(ing)" explored in this chapter might be imagined as a knot of ideas (i.e., of personhood, kinship, interspecific boundaries and bonds, livelihood, protection, etc.). Many of the ideational strands in this knot engage with conceptions not quite Dukha (i.e., my own ideas, and those which I garner and reappropriate within and beyond the borders of academia) while both making sense with and among Dukhas in the taiga, and technically describing the ontogenesis—the coming into being—of Dukha households. I render this ontogenesis as a processes along which Dukha households emerge in an ongoing manner and in circumstances of continually changing sorts of internally variegated changes.

As I conceptualize things here, household and householding are essentially the same subject-process. As actual phenomena, Dukha households are dynamic in the sense that they shift from place to place along migratory routes weaving together certain pastures, dwelling-places, and places of worship through dwelling, devotion, and passage over time, and in the sense that their compositions rhythmically change. The compositions of households are multidimensional, multidirectional, and "fractal" (Wagner 1991, 163). Households both compose and are composed. On the one hand, they co-compose broader ephemeral yet largely reoccurring settlement groups (i.e., aĭls and settlements composed of multiple aĭls), the taiga, and administrative units the concentrically scaled and centrifugally hierarchichized organization of which is exemplary of how "Ptolemy's scalar cosmography arguably persists today as an unexamined, taken-for-granted dimension of the concept of scale," according to which actions at local scales ought to be thought of, designed, and presided over at larger scales (Olwig 2011, 405).

On the other hand, households are composed of and by "persons," an ontological category conceived of here as "more-than-merely-human" (Donovan, Rasiulis, and Raycraft 2019). In terms of technical sustenance and achievements, household composition goes beyond only domains of humanity, and beyond only human-livestock relations, into realms of plants, minerals, crafts and market goods, and spirits. The household might be understood as a dynamic knot in the joining, disjoining, and rejoining of which is a locus of encounter all those realms; it is the site of ongoing performance through which chaos is partially and ephemerally tamed to serve human comfort, well-being, and even—for some—prosperity. Furthermore, householders are composed of different personalities, dispositions and, to some extent, species (i.e., traces of eaten food, bacteria in digestive tract, etc.). Along their relations, householders themselves compose broader groups and affiliations which might exceed the belonging to or bounds of any particular household—or of any scale and locus of fractal personhood, reaching from particular individuals and aĭls to and beyond the broader taiga.

In referring to the fractality of person(hood), I follow Roy Wagner's (1991) conceptualization of "the fractal person" (159), itself a re-entitlement of the "neither singular nor plural" person conceptualized by Marilyn Strathern under the referential guise—borrowed from Haraway (1985)—of the "cyborg." Wagner's (1991) re-entitlement aligns itself with "the mathematical notion of a dimensionality that cannot be expressed in whole numbers" (162). Speaking to "Benoit Mandelbrot's fractal dimensionality" as a "dimensional 'remainder' that replicates its figuration as part of the fabric of the field through all changes of scale" (Idem, 166), Wagner posits that fractality is a "holographic" configuration, a "self-scaling form [that] differs from a 'social organisation', or cultural ideology in that it is not imposed so as to order and organize, explain or interpret, a set of disparate elements. It is an instantiation of the elements themselves," or in other words: "a matter of the realisation of something that is already there, as the [materials] are already there" (Idem,

162). The fractal person "'keeps its scale,' as the mathematics of fractals would have it, regardless of the level of magnification" (Idem, 170). Integral here is "the fractality that prevents any differentiation of part from whole, that keeps the imageries of understanding from collapsing into the individuals, groups and categories that constructionism bundles into wholes greater than the sums of their parts" (Idem, 171).

In the context of the taiga, where does any person end and Oron Khangaĭ—the subjective personality and might of the taiga as a whole—begin? Looking to fractality as a framework for thinking with manifold person(hood)s in northernmost Mongolia might help us engage with such porous continuums of experience and identification. Further, our conceptual work with fractality might benefit from the nuanced attention to particularities—however broad the scale—emphasized in Strathern's (1992) conceptualization of "merographic relations" (72–73). Such relations are "made between parts of social life in a way that sustains the individuality of each for anything may be distinguishable as a part of something else, so nothing is ever simply part of a whole because another perspective, or interpretation, may redescribe it as part of something else" (Strathern 2018, 238). Looking to merographic relations helps us attend to the persistence and polymorphous porosity of individuality and particularity in and across all scales of fractality. Following Strathern further, we might look to "[c]ontextualizing, taking multiple viewpoints, switching perspectives" as devices of analytic modes that "sustain such [merographic] relations" (ibid.) in experiential and communicative engagement with fractal personhood among Dukhas.

Before proceeding with the account of the ontogenesis of Dukha households, the term "ontogenesis" should be fleshed out. Following Ingold (2017), the process of "ontogenesis—[. . .] the generation of being" (9) is understood here as the continual, immanent, improvisational, historically situated and generative, relational emergence of life and of what living beings come to experience. Ontogenesis emerges as respiration, the "rhythmic alternation" of two types of movements, one of "gathering or recollection," the other of "propulsion," associated with inhalation and exhalation, respectively (Ingold 2015, 66). Here media and materials are "dual aspects" of ontogenesis: "atmosphere on the inhalation, and meshwork on the exhalation" (Idem, 87). Living beings are the "site[s] where atmospheric immersion is transformed into the haptic extension of the meshwork along its proliferating lines" of becoming (Idem, 88). The sympathetic mutual responsiveness of meshwork and atmosphere, as well as the manner in which they bleed into one another or interpenetrate rather than interact as if they were bounded isolates, is exemplary of "correspondence," a co-constitutive movement of mutual attentiveness and adjustment in-between things (Ingold 2013, 107; Ingold 2015, 7, 11).

(LEIT)MOTIFS MOVING ALONG SCALES, DOMAINS, AND TEMPORALITY

In the following subsections I present a technical account of the ontogenesis of Dukha households. This account traces this variegated process throughout a conceptual continuum of encounter and entanglement amidst and in-between particular (merographically emergent) and broad (fractally emergent) constellations of both individuality and relationality. Engaged with in such a continuum, the household strikes us as a nexus and a crucible. It is a locus for the correspondence of beings, domains, and scales along modes of relation, the ongoing embodiment of which further generates and characterizes the composition and relational dynamics of each correspondent actor. The household perdures in its bridging of the particular and the broad, as well as of the local, the translocal and the alien.

Homeland: Ecological and Spiritual Landscape

The Dukha homeland is a meshwork, not only of diverse materials and living organisms, but also of ontologically and spiritually distinct places. Dukhas dwell along certain fibres in a web of sorts of places, namely lands suitable for inhabitation, uninhabitably sacred "ezenteĭ gazar" (Mngl.: "places inhabited by master spirits") or other places of personal or group worship, uninhabitably wicked "chötgörteĭ gazar" (Mngl.: "places haunted by demons"), and ambiguous spaces in-between (Kristensen 2004; Pedersen 2009; Kristensen 2015; Rasiulis 2016). Within the taiga's web, different places, spiritual forces, and semantic categories—each as labile as the other—are both distinct and, having no clear and constant delineations (Pedersen 2009, 143), porously open to one another in liminal zones. Drawing on a Dukha description of homeland as "boundless (xilgüi), and the political distinctions [. . .] merely names of the land (gazaryn ner), not boundaries in the land" (Idem, 142), Morten Axel Pedersen (2009) has eloquently exhibited how boundlessness characterizes not only the material and political dimensions and domains of the taiga, but also its spiritual ones.

This homeland, commonly referred to by Dukhas as "taĭga," the colloquial description of that which fits the literal definition of "Oron Khangaĭ" (Mngl.: "hilly, wooded, cool country / home plenty in water and good pastures"), a common proper name used in reference to and relation with the subjective, agentive ecology of their homeland. Dukha homeland is inextricable from ecology, spirituality, religiosity, and shamanic practices broadly speaking. Within this nexus of inextricability, there seems to be fractality. Oron Khangaĭ and each particular species, person, or place are as much fractals of the taiga as the latter is itself a fractal of local particularity.

As I have come to relate and know with them, the taiga, its inhabitants and its places are rugged, relentless in their perdurance while being continually exposed to wild weather, and ever changing as they interpenetrate one another and emerge along with the atmosphere they collectively and creatively undergo together as they fashion their variegated material becomings. Fractals resonate boundlessly throughout the taiga and across its inner frontiers like rhizomes characterized by a shared motif in all the movements of all that which or whom compose(s) it. Are there nonetheless leitmotifs which persist as distinct within this flood of relationally emergent fractal motifs?

I suggest that leitmotifs emerge and persist in the merographic relations that characterize the ontogenesis of the taiga. Here I draw on John Warrack's (1980) definition of Richard Wagner's technical notion of the "leitmotif" as "a theme or other coherent musical idea, clearly defined so as to retain its identity if modified on subsequent appearances," the purpose of which "is to represent or symbolize a person, object, place, idea, state of mind, supernatural force or any other ingredient in a dramatic work" (644). The leitmotifs I speak of and think with in this chapter are neither merely musical nor merely ideational, nor do they merely represent or symbolize; they meaningfully and materially are actual phenomena, "features of the world in their own right" (Pedersen 2011, 36). Individuality persists through the relationally entangled growth of living leitmotifs through the perdurance of "diffractive [. . .] differences that matter" (Barad 2007, 88). Such differences are what make leitmotifs distinct, yet they also constitute a major part of the contextual field which characterizes and to some extent structures all dramatically involved components' actions and relations. Drawing on Gregory Bateson (1972), such differences "bring about" the "effects" in the world (458). It is such difference that "gets onto the map[s]" (Idem, 457) Dukhas make of their homeland—and of their "asjigtai [(i.e. aszhigtaĭ)]" (Kristensen 2004), the ways humans ought to live in relation with myriad differently disposed beings and groups with their own distinct aszhigtaĭ in this shared homeland—in ongoing negotiation and exploration with their peers (ibid; Kristensen 2015; Rasiulis 2016).

Effectively inhabiting the political, ontological, and spiritual porosity and spatial uncertainty of different places among and amidst different sorts of beings and groups in the taiga requires both semantic and practical flexibility, as well as skillful, sensible, knowledgeable, improvisational movement that helps position oneself, one's kin in sympathetic and even reciprocal relation with other humans, animals, and spiritual beings or forces. Ergo, Dukhas engage the taiga a subjective, ambiguously labile neighborhood (Kristensen 2004) with whom and within which they dwell, subsist, identify, and engage in spiritual practices. Dukhas coordinate their engagement with this territory by inhabiting it in ways which position them as non-exclusive "masters of the land" (Kristensen 2015, 62) who steward, or "kharakh" (Mngl.: "watch over"), their homeland's well-being as it sympathetically corresponds with the realization of their livelihoods.

Realizing their livelihoods in sympathy with spirits directly or indirectly perceived in the landscape, Dukhas cooperatively orchestrate the impacts their use and worship of their "boundless" (Pedersen 2009, 141) homeland have on its ecological and spiritual well-being (Kristensen 2015, 147). Recognizing diverse spirits in lands riddled with ambiguous and labile frontiers, and acknowledging some of those beings as greater, potentially cooperative land masters (Idem, 86), Dukhas actively participate in a process of homeland stewardship in which humans are not the only actors. Among these actors, the agent nearest in bearing and activity to the spirit and function of a land management institution might be Oron Khangaĭ, the supreme master of the taiga, by whom its inhabitants are protected and nurtured if and when they live according to the master's "yus" (Mngl.: "rule/s") rules of interspecifically non-obstructive conduct (Idem, 182, 199). Yet, within the gaze and composition of this supreme master, the manifold beings of the taiga themselves strive for mastery in and along their own distinct, dynamic niches.

Drawing on Donatas Brandišauskas' (2009) work among Orochen reindeer pastoralists and hunters in the nearby Baikal region, we might understand "mastery" as the "adjusting of one's own dwelling activities to the shifting environment as seasons or animals movements as well as to places linked to master-spirits, so that one can be in the right place and be ready for different opportunities in subsistence," thereby masterfully "maintaining luck and wellbeing [. . .] while cooperating with non-humans or achieving success through aggression in the environment of contest" (254). Mastery is as paramount as ever to life in the taiga at scales in which alien actors and forces mingle or meddle with that and those which and whom compose the taiga.

DWELLING AND MASTERY IN AN ATMOSPHERE OF HISTORY, POLITICS, ECONOMY, AND MANIFOLD MIGHT(S)

Most present day Dukhas are descendants of those earliest of inhabitants of Eastern Tyva who eventually became known under the appellation "Todjins," as identified by Sevyan Vainshtein (1980, 121), or "Tozhu," as identified by Brian Donahoe (2002). All present day Dukha reindeer pastoralists technically embody the same "economic-cultural type" (ECT) as did seventeenth to twentieth-century Tozhu and their predecessors arguably dating back to circa AD 500 (Humphrey 1980, 9; Vainshtein 1980), which Vainshtein categorized as the "Sayan upland taiga sub-type of the hunters and reindeer-breeders of the Siberian taiga" (Idem, 49; original emphasis). Of course, current embodiments of this ECT are characterized by distinct adjustments to traditional practices, and by the emergence of new socio-economic strategies and synergies in correspondence with the ongoing historicity of the taiga and of Inner Asia.

A reading of Vainshtein's (1980) Nomads of South Siberia: the Pastoral Economies of Tuva helps us grasp how, throughout epochs of successively Scythian, Hun, Tugyu-Turks, Uigur, Kyrgyz, Mongol, Manchu, and Imperial Russian influence, Tyva and its distinct yet deeply entangled peoples have been largely autonomous, as well as economically, technologically, and politically versatile in their position(s) at the frontier(s) of geopolitical frontiers and borderlands while continually engaged in relations and processes seeping far beyond the taiga and even Tyva. Perhaps the most extreme frontier had always been Eastern Tyva, where the communities of hunters and reindeer breeders, by virtue of their dwelling in remote, biodiverse areas rich in valuable minerals, had the need, means, knowledge, and technical proficiency to be extremely autonomous and versatile (see Figure 7.1).

Although they traded for various items or materials from beyond the taiga and at times Tyva, Tozhu lifestyles were so unique that many of the tools and objects they used were only made locally, in many cases within each

Figure 7.1 Reindeer and Horses of Different Households. Some are Let Loose, Others Tied Up. *Source*: 2018, N. Rasiulis.

household or aĭl. Arguably, the input of alien resources and technologies into the upland taiga of Todja was far outweighed by the output of resources, notably market-standard pelts (i.e., squirrel, sable), specialty pelts (i.e., fox, otter, bear, wolverine, lynx, etc.), and, we might surmise, marsh salt, antlers and horns, and precious metals, and gemstones. The main output of the area consisted of pelts, and arguably the majority of this output was to satisfy the quotas of tributes paid to various administrators, fluctuating periodically from local to alien, at the (furthest reaches and slightest efficacy of the) bidding of different political entities over time. With such heavy outputs of key resources, Tozhu and indeed all Tuvinians were integral to the growth and strength of powerful alien political entities and representative figures of these entities who, at a frontier of these polities, were largely absorbed into what steadfastly remains an economically and culturally distinct Tyva-Shishged drainage system area.

That some Tozhu households succeeded in moving throughout the furthest reaches of any government territorialization in Inner Asia during the 1920s and 1950s, as they evaded repetitive militarily enabled campaigns in the Soviet Union and in Mongolia to seize or "to expel" them by means of "forced removals" (Wheeler 2000, 41), is a testament to the situated abilities of those households and aĭls. These groups were able to co-habit with sufficient livestock (especially reindeer steeds and beasts of burden) for extensive shifting in the furthest of frontiers, and/or with sufficiently strong social relations throughout broader networks of livestock-rich households and aĭls to have access to enough livestock for such remote migrations. Dwelling in the furthest of frontiers enabled those households and aĭls to autonomously make a living in the taiga in active correspondence with outside forces and factors throughout times of violent upheaval. That Dukha reindeer and horse husbandry-enabled hunting and gathering lifestyles perdured throughout the socialist and post-socialist epochs each characterized by contradictory coincidences of benefits and traumatic upheavals is a testament to the ongoing proficiency of Dukha households, aĭls, and broader relational networks of kin and friends to improvise a living in the taiga amidst unpredictable changes in situations brought forth largely by outside forces, to which local actors continually adjust.

During the 1990s and early 2000s, as the service, provisioning, and salary infrastructures of the communist Mongolian state withered away and finally collapsed, many Dukha householders returned to the taiga and to reindeer herding and hunting and gathering lifestyles and economic pursuits after several decades of scanty or merely peripheral ties to the taiga and its particular correspondent dwelling practices. As Benedikte Kristensen (2015) has shown, the years of absence of many Dukhas from the taiga allowed neglect of ongoing personal and kin relations with various spirits to fester. As a result, the taiga and its resident spirits were particularly volatile in relation with Dukhas who, for the most part, were uneasily figuring out for themselves after a long hiatus—or for the first time—how to act appropriately in different places and in relation with diverse spirits in the taiga (235). It is through further (re)immersion in the traditional livelihood practices of hunting, herding, gathering, and nomadizing that Dukhas directly engage with their homeland in an ongoing fashion, and it is by these means only that one learns how to effectively and appropriately move and relate in the taiga. It is also by these means that one further exposes oneself to uncertainties of both physical and spiritual tenors.

Following Pedersen (2011), we might conceive of post-socialism in the Shishged area as a state of "ontological meltdown [. . .], this slow but relentless breakdown of the world as people had known it" (8), which, for local people was not "merely a transition from one political and economic system to another," but rather the exposure "to a violent intrusion of invisible forces, energies, and substances, which, during seventy years of socialism, had hovered only in the shadowy margins of self, household, community, and nation" (ibid). In such a chaotic, uncertain atmosphere, "agency above all was in the hands of persons imbued with the shamanic capacity to change in accordance with the changes of transition itself, while at the same time retaining a semblance of form" (Idem, 223). Pedersen argues that "the potential shamans, and they alone, are imbued with a hyper-shamanic capacity to mold, extend, and apportion emergent social, political, and economic realities into new impossible forms, making it possible to ride on the edge of change" (Idem, 224). Whether or not these potential or "not quite shamans" (Idem, 222) are the only persons disposed in a manner which allows them to "ride on the edge of change," all inhabitants of the Shishged, none the least Dukhas, continually undergo "the edge of change," and, lest they be swept away into oblivion, manage to improvise a way forward in our whirligig of a world with other persons—notably through the leitmotifs of households and aĭls.

The issues and dynamics brought forth or (re)emphasized by post-socialism in the Shishged are further complicated and compounded by the establishment and seemingly haphazard implementation of a national park, as well as by the freeform, highly competitive and immensely diverse tourism industry it espouses, leeches onto, and empowers. Since 2011 the slice of the Saian homeland which is currently Dukha territory has been legally and practically subsumed under the rubric of a state-run environmentally protected area (Johnsen et al. 2012, 23) entitled the "Tengis-Shishged National Park" (TSNP). This "sister-park" of Yosemite National Park (US National Park Service 2016) seemingly ignores the practical and ontological realities of Dukhas' territorial engagements by applying a transnationally propagated, universalist model of "top-down conservation" (Mulrennan, Rodney, and

Scott 2012, 244) according to which territory is subject to centralized management which forcefully regulates human activity with the aim of protecting the biodiversity of a seemingly meek ecosystem supposedly devoid of agency.

Dukhas strategically adjust their practical use and worship of their homeland to changing ecological conditions in a fashion which allows ongoing processes of symbiosis within and beyond domestic constellations to go on in and along continually changing ways. Doing so, Dukhas develop "situated" (Nazarea 2006, 324), integrative abilities and knowledge of primal salience in their territory. However, TSNP management does not inform or nuance its operations with these situated aptitudes. This lack of collaboration engenders frictions such as those that arise from and in response to many if not most—and, for some people, all—tenets of TSNP regulatory framework (i.e. restrictions on mobility, hunting, fishing, access to pasture, etc.). The "(corpo)realization" (Blaser 2010) of these tenets hinder Dukhas' legal and effectual realization of their (arguably) sustainable livelihoods and ongoing stewardship—enmeshed with those livelihoods—of their homeland (Rasiulis 2016, 101). Furthermore, TSNP tenets threaten ecosystem health if only with situationally inescapable overconsumption of resources in those relatively small, constraining areas allocated by the park for human inhabitation (Rasiulis 2016, 26, 101).

As its purpose is not limited to environmental protection, the TSNP is not merely a state apparatus for nature conservation. The existence—and internationally recognized statuses (i.e., International Union for the Conservation of Nature (IUCN) Category II protected area, US National Parks Service sister park, etc.) of this national park generates "bureaucratic capital" (Heikkinen et al. 2010, 122) for the province of Khovsgol, the Mongolian state, and, following Jim Igoe (2010), transnational non-profit or for-profit enterprise entities in a globalizing atmosphere increasingly neoliberal in tenor, in which nature conservation is increasingly not only analyzed in financial terms but also monetized in the creation of new, experience- and social media content-based commodities. In this manner, the TSNP might be understood as an institution that reproduces what Elizabeth Garland (2008) has called the "conservationist mode of production" (62; in Igoe 2010), one that adds economic value to natural capital through various mediations which transform it into a convertible kind of global ramification (Igoe 2010, 376).

Living within the grasp of the TSNP's land-use proscriptions and prescriptions, and of the "conservationist mode of production" it enables—yet continually seeping beyond its grip—Dukhas are no longer required by environing political entities to provide resources such as pelts that might then be transformed into commodities. Rather, Dukhas are now financially supported by the state with monthly stipends, dispersed to each householder of all households who have registered possession of reindeer with the government, as compensation for the hardships brought about by lack of consultation concerning environmental protection regulations, and by corresponding incongruities between Dukha—and indeed taiga—lifestyles and the TSNP regulatory framework. We might speculate that this state-subsidized reindeer husbandry allows state and non-state actors to indirectly engineer—or rather partially structure and condition—Dukhas and their lifestyles into a commodity for touristic consumption and "prosumption"—the blurring of consumption and production characteristic of conservation in the present era of highly interactive and largely user-generated internet content—(Büscher and Igoe 2013, 284), and for the legitimatization of the TSNP's status as an IUCN Category II protected area.

The engineering of Dukha persons, households, aĭls, private and collective herds, pastures, and indeed all their broader taiga networks into symbols and shells of a unified "wilderness—primitive pastoralist" commodity involves "spectacular accumulation" (Tsing 2005; cf. Igoe 2010; Igoe 2017), which Igoe (2010) has identified as rampant in capitalist-conservationist joint ventures such as IUCN recognized national parks. "Spectacular accumulation" is a mechanism through which certain conservation projects garner support and means as well as engender material outcomes by circulating imagistic and narrative curation that obscures or outright occults certain facets of reality while emphasizing others within a discursive framework which feigns not only cohesiveness but also comprehensive accuracy (Idem, 380, 386). Ideas of desired forms of nature and of human interactions with nature are created. Endeavors spun as ideally disposed to engender such forms are facilitated with influxes of capital investment or donations as well as with popular support in public opinion brought about in the dissemination of sensationalized images and narratives regarding those desired forms and the socially ecologically ethical ventures through which interaction with or adoration of these forms is possible.

Through the "Blue Waves Campaign," the Rally for Rangers Foundation—a transnational charitable organization sponsored by various commodity brand and media corporations—has worked in partnership with the Mongol Ecology Center, a transnational non-profit NGO, to raise funds to equip TSNP park rangers with high grade all-terrain motorcycles with which the park could be more effectively surveyed, making it easier to arrest poachers (Rally For Rangers Foundation 2017; Mongol Ecology Center 2014). As in the numerous cases of prosumption in conservation illustrated by Igoe (2010, 381–382), the Blue Waves Campaign employs spectacular accumulation, empowered by the involvement of celebrity spokespeople (i.e., famous motorcycle riders), to present photo and video fragments as a coherent and comprehensive portrayal of an easily understandable and solvable conflict between poachers, on the one hand, and rangers, on the other hand, the latter fighting to protect global biodiversity, from which far-away prosumers benefit, in local places for the good of happy, resilient remnants of an ancient—in this case, shamanic—culture almost as inspiring as the wilderness they inhabit.

In the Blue Waves Campaign, and in the broader engineering of a "wilderness—primitive pastoralist" commodity in the TSNP to which that campaign contributes, the main facets that are emphasized are the awe-inspiring magnificence and anthropocentrically valued, globally significant biodiversity of local landscapes, as well as the "noble savagery" of Dukha households—conical tent and all—and lifestyles associated with oversimplified, imaginatively glorious shamanism and reindeer husbandry. Obscured or occulted are not only the complexity of Dukha animal husbandry and shamanism, but also many Dukha traditional livelihood practices (i.e., hunting, gathering, ezen worship, trading minerals, wild animal and plant materials, etc.), the technical means through which social-ecological integrity is maintained through lifestyles that attend to the radical relationality of the taiga. Occulted here are also the contradictions inherent in criminalizing some of the most important facets of Dukha livelihoods, and those through which the spectacularly emphasized and administratively prioritized practices of reindeer husbandry and tourist-friendly shamanism have largely been sustained throughout history.

I propose that we might conceive of what seems to be an emerging "wilderness—primitive pastoralist" commodity as an "impossible commodity." Here I draw on Pedersen's (2011) notion of "impossible forms" (34), which he uses in reference to the "labile shapes" with which "the invisible powers of the spirits and the opaque forces of the market are imbued" (ibid). The "wilderness—primitive pastoralist" commodity is impossible insofar as it is labile and entangled with similarly labile invisible forces emanating and radiating both in and beyond the taiga, and also, departing slightly from Pedersen's use of "impossible," insofar as it is largely an imagination that is untenable in practice yet that bears practical consequences. As living leitmotifs knotted into and out of the very fabric of the taiga, persons, households, aĭls and the like cannot be merely commodities. Further, the marketed qualities of the "wilderness—primitive pastoralist" commodity are unattainable and unsustainable without the perdurance of markedly outlawed—and thereby materially restricted—practices, and of dimensions of the taiga that are more complicated and grave than most aliens might imagine.

Yet, laws are enforced, subsistence-hunting becomes in most instances either illegal or legal only with the purchase of special permits that arguably no Dukha can afford. Furthermore, according to TSNP regulations, dogs—by virtue of their usefulness in trailing prey—are not permitted in most traditional hunting grounds. To cope with this de facto outright ban on subsistence-hunting, Dukha householders and households adjust their lifestyles as they increasingly depend on cash income for the purchase of domestic meat from nearby steppe-dwellers. Hordes of tourists, tour guides, and tour operator enterprises flood the taiga and Dukhas every year, largely ignorant of the problematic issues with which this sensational ecotourism destination and its hosts are wrestling. In this manner, a Dukha man—who happens to be a shaman—can be seen riding a white reindeer through a snowy taiga landscape in the image on a postcard model on sale at the Mongolian central post office; this postcard can occlude the fact that the man's son was recently arrested and charged with "poaching" while allegedly out with a subsistence hunting party (Gauthier and Pravettoni 2016); and this postcard can occult the true character(s) and inextricability of Dukha reindeer husbandry, hunting, and shamanism, as well as the perdurance of householding beyond the postcard and despite householders risking or indeed finding themselves behind bars.

Ironically, the criminalization of life at the underbelly and in the farthest reaches of the Mongolian taiga increases the financial value of resources which can only be secured in the practice of activities prohibited by the Mongolian state within the contested boundaries of the TSNP. This criminalization can even create new or newly growing forms of profitable delinquency in the area, as was evidenced in the emergence of prostitution rings during the illegal gold rush, circa 2010 (Kristensen 2015, 138), and as is still evidenced in the ongoing clandestine sale of store-bought alcohol at inflated costs in the taiga, further complicating the endeavor of environmental protection within the TSNP. Perhaps nature conservation—at least as it stands in northernmost Mongolia—might itself be conceived of as an "impossible endeavor."

HOUSEHOLD(ING): HOLDING FAMILY AND HOME TOGETHER IN(-BETWEEN) PLACE(S) AND TIME(S)

Companionship, hospitality, and teamwork are necessary for thriving both as individuals and as communities in the rugged, complex, task-filled environment of the taiga. Here there is a near or outright inextricability of householder, household, aĭl, and broader networks through which Dukhas continually disperse and diversify their economic endeavors both in and out of the taiga. Here there is also a technical proximity of householding and nomadic modes of radically relational, acrobatically improvisational livelihood realization. Here, in maintaining both household and herd, "[d]omestication is not something that is finished, and neither is it fixed. It is not a result;

it is a process through which the proximity of humans and animals constantly changes" (Davydov 2014, 9).

Dukhas orchestrate their complex itinerancy in ways which mediate their rhythmically fluctuating ecological context with the various needs of diverse livestock species, relationships with friends and family, to cellular reception, to places of religious worship, and to tourism-based income opportunities, as well as access to the village which serves as the administrative and market center of Tsagaannuur Sum. Albeit the diversity of factors experimentally taken into account by Dukhas, their movements necessarily accommodate the particularities of their reindeer with particular structurational emphasis. This is a testament to the integral roles these animals play in the ongoing growth of Dukha livelihoods and households.

In my experience, Dukhas effectively realize livelihood practices in a cheerful, albeit sober and nonchalant way. They skillfully anticipate and perform diverse tasks, each involving skillful discernment and movement. Tasks are carried out in a playful and cooperative way, together with friends, family, and other animals along nomadic ways of life in a wilderness habitat that is navigated under the leitmotif of the household. Yet this very life continually seeps through the porous frontiers of the household. Dukha lifestyles catalyze the development of poised improvisation and acrobatics, or

Figure 7.2 Man Leads Pack Reindeer Along Path Too Steep to Ride Down. His Niece Herds their Aĭl's Horses. *Source*: 2018, N. Rasiulis.

"freestyle bearing[s]" (Rasiulis 2016). Thus they rely on situated knowledge with which such bearings are informed and thereby rendered all the more effective.

As taiga-based livelihoods include boundless, seamlessly interwoven tasks at any and all times, the practice of said livelihoods is undistinguishable from the everyday experience of inhabitation or dwelling itself. Throughout their entire lives, Dukha people learn and increasingly enskill themselves as they resourcefully engage diverse materials, effectively engendering anticipatively intended goals. The physical traces of these enskilling performances are left in the land, and they incarnate the stories of their occurrences as monuments of past inhabitation. These monuments are potentially meaningful to those who attend to them with their senses, memory, and imagination. Following Brandišauskas (2009), these performative acts and their traces contribute to rendering certain places inhabitable, in many instances again and anew (254), always at the frontiers of domesticity and change in the taiga.

As they inhabit the storied taiga, Dukha householders embody freestyle bearing in practices of making things and homes, of moving around in the taiga, and of stewarding multispecies herds, all the while navigating swells of emerging market, black market, and regulatory dynamics. The masterfully mighty, quick-witted and materially sensitive acrobatic improvisation of Dukha lifestyles makes them quite resilient, capable of somehow maintaining an evolving tradition of reindeer husbandry—which, stemming from the origins of Saian reindeer breeding dates back to "at least [. . .] the first century A.D." (Vainshtein 1980, 120). It also positions Dukha reindeer breeding—arguably "the oldest in Eurasia" (Idem, 121)—in a quickly changing social, economic, technological, and geo-political context.

Brandišauskas (2009) notices a similar improvisational mode of ongoing environmentally "response-able" (van Dooren and Rose 2016, 89) adjustment among Orochen herders and hunters. He describes the experience of weather as inextricability tied to the lifestyles of the "Sayan upland taiga sub-type" of Siberian hunting and reindeer breeding, a mode of living I have elsewhere formulated as "open-air nomadism" (Rasiulis 2016, 16). In this dwelling mode, weather is never outright separate from the sanctity of domestic space or practice. In Brandišauskas' (2009) words, "mastery [. . .] is linked to one's predictions of weather, adjusting own movements to weather shift[s] [. . .] through 'reading changes' and relying on own embodied experiences [. . .] while sharing" experiential knowledge with others in cooperative fashion (257).

We might extend this notion of weather-based anticipation and adjustment to the entire economic, political, and socio-spiritual atmosphere, or what Irving Hallowell (1988) has called the "total 'behavioral environment'" (87; in Santo and Blanes 2014, 25), the whole labile field of chaotic activity permeating or invading the taiga. As it is with weather, so too it is with all atmospheric facets of householding, none the least those that correspond with hunting and with remaining free from behind bars: "technical know-how is intertwined and fused with spiritual know-how, and the two domains of knowledge are not conceptually distinguished" (Willerslev 2007, 134).

Among Dukhas and elsewhere in Siberia, the technical means through which "one's aims to secure mastery of places" are embodied "not only through rituals led by shamans, but also through daily strategies, skills and knowledge of subsistence" (Brandišauskas 2009, 259). Realizing these strategies, people draw on "informal networks of cooperation while aiming to achieve luck or gain [. . .] success" (ibid.) in diverse yet mutually integral economic pursuits. In the current atmosphere of the taiga at the manifold frontier of state territorialization in northernmost Mongolia and of public-private joint ventures that embody a capitalist mode of (commodity and territory of commodities) production, "herders and hunters [. . .] rely on accumulation of goods and autonomy in [the] taiga as well as [on] covert and overt strategies securing their subsistence" (Brandišauskas 2009, 258). Here, as it is in the Baikal area, "interactions are based on complex relations as well as experiences infused with anxiety, ambiguity and doubt," and "subsistence is not simply dwelling with the flow of environment" (Idem 260). Rather, "life is lived in between and at once both mastering over it and mastering together with it" (Brandišauskas 2009, 260), compelling us to acknowledge that secrecy, "aggression and cooperation can be seen as part of reciprocity," (Idem, 253) of mastery, and of might.

We might conceive of Dukha householding as a labile, variegated form of mastery through which personal and collective leitmotifs entangled in radical relationality are sustained and grown with might along astutely improvised movements. This type of householding is both fractal and merographic in that it is atonce harmonious and conflictual. Here the coexistence of subversion and yielding affords cooperation in situations where power dynamics between different sorts of beings are asymmetrical, by virtue of those beings' intrinsic qualities and/or of an arguably (historically) (en)forced cohabitation. Integral to the household's co-composition, and thereby deeply implicated in contradictory cooperation, are not only the human householders, but also their livestock, dogs, wild animals, plants—notably sacred plants hung in tents, as well as pastures that feed livestock and sustain pastoralist lifestyles—and other beings, namely the guardian spirit that inhabits the hearth whence, if suitably fed and cared for, it watches over the household's well-being.

CONCLUSION

Migrations are momentous performances on the part of Dukha households' mammal composers. Migratory performers attend to the needs, characters, and dispositions of each of the household's constituent species and individuals—none the least the humans (arguably) orchestrating the whole shebang—on-the-go. Here migratory routes are understood not only in terms of broad distribution and timing of camps and shifting between different camps and corresponding pastures and clandestine hunting grounds, but also in terms of particular ways and paths fared from one place to another and from one sort of domain to another, over time, and among other households and aĭls. Exemplary of the latter dimension of migratory routes, travelling from a summer camp to a fall camp in 2018, Dukha mentors and I were moving away from flatter paths that could have been fared more quickly on horseback. The reasoning for this choice of movement was manifold. The eldest composer of our aĭl had chosen to ride a reindeer, and the slower, more mountainous, forested, lichen-rich, and windy way was more accommodating to his ride. Furthermore, traveling in caravans, with some aĭl composers steering their collective herd of horses toward our destination(s), required a loose synchronization of the speeds of steeds and unridden horses in the herd. This synchronization was facilitated by the layout and composition of the lands in which our slower-speed migration fared. The chosen way also accommodated the dogs who live peripherally to—and to varying degrees within—the aĭl's households. In the lands fared, these dogs were jogging and juking with ease and delight at our synchronized tempo. Such a slower tempo results in the longer duration of migratory trips. Although managing livestock during migrations is more difficult than when in settlements, it is facilitated by the makeup of the lands fared.

Over time, migratory pathways are trodden by many persons, households, and ails, as is evidenced in the many 'ovoo' cairns atop Buhairlagiin Dawa mountain pass in West Taiga where our caravans passed in 2018. Each ovoo had been generated over time by different constellations of householders, households, and clans. The cairns remind us that household are composed of more than merely humans, animals, and, arguably, vegetation. Households are configured to include spirits, rocks, and various nonhuman or more-than-human others that and who perdure with might and mastery. It is debatable where we might draw the line (or not) between entities who are either, on the one side, composers or, on the other, components of household configurations and aĭl constellations. Yet it is arguable that every person and/or thing of whom or which these configurations and constellations are composed are enmeshed in particular, perdurant, and strong ties of conviviality and of mutual (re)generation that permeate across domains of life, matter, spirit, and might in the "très longue durée" (Bergson 2002 [1907], 212). Here, reindeer and indeed all livestock are not mere property, tools, or statistics, but rather beings with their own unique, valuable characteristics as "partners" in ongoing relations through movement. They are continually getting closer to and further from frontiers between humans (Davydov 2014, 10) and persons of other species or ontological domains (i.e., spirits, nature conservation actors, etc.).

Although Dukhas perceive their reindeer as distinct from these animals' zerlig (Mngl.: "wild") counterparts by virtue of the former's condition as mal (Mngl.: "livestock"), and of the relative relinquishment of freedoms which engender such a condition, their herds of forest reindeer are categorized by scientists as "semi-domesticated" (Johnsen et al. 2012; Küçüküstel, 2013, 62). Rather than fully subjugated to human control, the reindeer playfully dance in and out of the fluid boundaries of human authority. Discussing Tozhu reindeer pastoralists, Stépanoff (2012) posits that in order for reindeer, who remain free to escape humans and their encampments, to willfully decide to come back toward camps where they are herded, caught, led, and ridden by humans, herders must mediate practices of dominating reindeer with conditions desirable to these ungulates (190). Likewise, Dukhas must temper their control over reindeer—a subversion of these animals' wildness—with attractiveness, and by periodically yielding to their needs, dispositions, and uncontainable freedom.

In the practice of everyday life, along with Dukha people in their nomadic inhabitation of the taiga, I was continually exposed to tangible examples of intimate human participation in symbiosis. I experientially grasped human participation in interspecific symbiosis as a playfully and painstakingly earned relationship of co-habitational cooperation in sustaining one another's lives, regardless of the measures taken by a symbiont or its interspecific counterpart to either yield to or subvert the other.

I now conceive of symbiosis as a co-habitational process in which a reindeer—whose body and character I know well, and who knows my body and character perhaps even more acutely than I know theirs'—prances over to where I will urinate as soon as they see me unbuckle my pant belt. Symbiosis is also sharing reindeer meat with a dog that has helped you hunt and has kept wolves at bay for years. Symbiosis is also killing your horse, after years of protecting it from wolves, before it grows too old and weak and falls to predation or sickness. You give it an honorable, swift death, and you eat its preciously grown meat, sharing it with friends, family, and even a foreign visitor. Symbiosis is co-habitation that stitches together interspecific frontiers in the interstices of which persist the fractality of broad phenomena. It sorts along merographic relations that sustain the steadfast particularity of different actors and domains. My own participation in the ongoing, symbiotic ontogenesis of Dukha households is a testament to the flexibility and dynamism of Dukha huseholding in a continually changing world-drama of increasingly diverse and shifting casts.

REFERENCES

Barad, Karen. 2007. *Meeting the Universe Halfway*. Durham: Duke University Press.

Bateson, Gregory. 1972. *Steps to an Ecology of Mind: Collected Essays in Anthropology, Psychiatry, Evolution, and Epistemology*. Northvale: Jason Aronson Inc.

Bergson, Henri. 2002 [1907]. *L'évolution créatrice*. Digital Version Produced by Gemma Paquet, Part of "Les classiques des sciences sociales" Series. Chicoutimi: CEGEP de Chicoutimi in Partnership with Université du Québec à Chicoutimi. Originally published in Paris: Les Presses universitaires de France.

Billé, Franck, Grégory Delaplace, and Caroline Humphrey, eds. 2012. *Frontier Encounters: Knowledge and Practice at the Russian, Chinese and Mongolian Border*. Open Book Publishing.

Blaser, Mario. 2010. *Storytelling Globalization from the Chaco and Beyond*. Durham: Duke University Press.

Brandišauskas, Donatas. 2009. "Leaving Footprints in the Taiga Enacted and Emplaced Power and Luck among Orochen-Evenki of the Zabaikal Region in East Siberia." PhD diss., University Of Aberdeen.

Büscher, Bram, and Jim Igoe. 2013. "'Prosuming' Conservation? Web 2.0, Nature and the Intensification of Value-Producing Labour in Late Capitalism." *Journal of Consumer Culture* 13 (3): 283–305. DOI: 10.1177/1469540513482691.

Davydov, Vladimir. 2014. "Coming Back to the Same Places: The Ethnography of Human-Reindeer Relations in the Northern Baikal Region." *Journal of Ethnology and Folkloristics* 8 (2): 7–32.

Delaplace, Grégory. 2012. "A Slightly Complicated Door: The Ethnography and Conceptualisation of North Asian Borders." In *Frontier Encounters: Knowledge and Practice at the Russian, Chinese and Mongolian Border*, edited by Franck Billé, Grégory Delaplace, and Caroline Humphrey, 1–18. Open Book Publishing.

Donahoe, Brian. 2002. "Hey, You! Get Offa My Taiga! Comparing the Sense of Property Rights Among the Tofa and Tozhu-Tyva." Max Planck Institute for Social Anthropology Working Paper Series 38.

Donovan, Amy, Nicolas Rasiulis, and Justin Raycraft. 2019. "Experimental Anthropologies for/with the More-than-Merely-Human." Paper Presentation Panel (refused), *11th Annual McGill Anthropology Graduate Conference,* Montréal, April 4–6.

Garland, Elizabeth. 2008. "The Elephant in the Room: Confronting the Colonial Character of Wildlife Conservation in Africa." *African Studies Review* 51 (3): 51–74.

Gauthier, Marine, and Riccardo Pravettoni. 2016. "'We Have Nothing but our Reindeer': Conservation Threatens Ruination for Mongolia's Dukha." *The Guardian.*

Published online August 28. URL: https://www.theguardian.com/global-development/2016/aug/28/reindeer-conservation-threatens-ruination-mongolia-dukha.

Hallowell, Irving. 1988 [1955]. *Culture and Experience*. Long Grove: Waveland Press.

Haraway, Donna 1985. "A Manifesto for Cyborgs: Science, Technology, and Socialist Feminism in the 1980s." *Socialist Review* 15 (2): 65–107.

Heikkinen, Hannu, I., Simmo Sarkki, Jokinen Mikko, and David, E. Fornander. 2010. "Global Area Conservation Ideals versus the Local Realities of Reindeer Herding in Northernmost Finland." *International Journal of Business and Globalisation* 4 (2): 110–130.

Humphrey, Caroline. 1980. "Editor's Introduction." In *Nomads of South Siberia: The Pastoral Economies of Tuva,* Sevyan Vainshtein, 1–36. New York: Cambridge University Press.

Humphrey, Caroline, and David Sneath. 1999. *The End of Nomadism? Society, State, and the Environment in Inner Asia.* Durham: Duke University Press.

Igoe, Jim. 2010. "The Spectacle of Nature in the Global Economy of Appearances: Anthropological Engagements with the Spectacular Mediations of Transnational Conservation." *Critique of Anthropology* 30 (4): 375–397. DOI: 10.1177/0308275X10372468.

———. 2017. *The Nature of Spectacle: On Images, Money, and Conserving Capitalism.* Tucson: The University of Arizona Press.

Ingold. Tim. 2013. "Prospect." In *Biosocial Becomings: Integrating Social and Biological Anthropology,* edited by Ingold, Tim, Gisli Palsson, 1–21. Cambridge: Cambridge University Press. http://dx.doi.org/10.1017/CBO9781139198394.002.

———. 2015. *The Life of Lines*. New York: Routledge.

———. 2017. *Correspondences*. In "Knowing from the Inside" series of special volumes, curated by Tim Ingold, funded through a European Research Council Advanced Grant held by Tim Ingold. Aberdeen: University of Aberdeen.

———. 2018. *Anthropology: Why It Matters*. Cambridge: Polity Press.

IUCN. ND. "Category II: National Park." IUCN website. Accessed online, July 10, 2019. URL: https://www.iucn.org/theme/protected-areas/about/protected-areas-categories/category-ii-national-park.

Johnsen, K. I., B. Alfthan, P. Tsogsaikhan, and S. D. Mathiesen, eds. 2012. "Changing Taiga: Challenges to Mongolian Reindeer Husbandry." Portraits of Transition No. 1. United Nations Environment Programme, GRID-Arendal.

Khazanov, Anatoly M. 1998. "Pastoralists in the Contemporary World: The Problem of Survival." In *Changing Nomads in a Changing World,* edited by Joseph Ginat, and Anatoly M. Khazanov, 7–23. Portland: Sussex Academic Press.

Kristensen, Benedikte. 2004. "The Living Landscape of Knowledge: An Analysis of Shamanism among the Duha Tuvinians of Northern Mongolia." *Specialeafhandling til Kandidateksamen* 317: NP. Institut for Antropologi, Københavns Universitet. http://www.anthrobase.com/Txt/K/Kristensen_B_02.htm. Accessed December 10, 2013.

———. 2015. "Returning to the Forest: Shamanism, Landscape and History among the Duha of Northern Mongolia." PhD diss., University Of Copenhagen.

Krupnik, Igor. 1998. "Understanding Reindeer Pastoralism in Modern Siberia: Ecological Continuity Versus State Engineering." In *Changing Nomads in a Changing World,* edited by Joseph Ginat, and Anatoly M. Khazanov, 223–242. Portland: Sussex Academic Press.

Küçüküstel, Selcen. 2013. "The Dukha and How They Perceive the Environment–Maintaining Good Relations with Land Spirits." MA thesis, Yeditepe University, Istanbul.

Lancaster, William, and Fidelity Lancaster. 1998. "Who Are These Nomads? What Do They Do? Continuous Change or Changing Continuities?" In *Changing Nomads in a Changing World,* edited by Joseph Ginat, and Anatoly M. Khazanov, 24–37. Portland: Sussex Academic Press.

Lemonier, Pierre. 2010. "Technology." In *The Routledge Encyclopedia of Social and Cultural Anthropology* (2nd edition), edited by Alan Barnard, and Jonathan Spencer, 684–688. London and New York: Routledge.

Mongol Ecology Center. 2014. "Sons of Taiga." https://www.youtube.com/watch?v=mQG8Xg29QW8&t=8s. Video accessed online January 7, 2018.

Mulrennan, Monica E., Mark Rodney, and Colin H. Scott. 2012. "Revamping Community-Based Conservation Through Participatory Research." *The Canadian Geographer/Le Géographe Canadien* 56 (2): 243–259.

Nazarea, Virginia. 2006. "Local Knowledge and Memory in Biodiversity Conservation." *The Annual Review of Anthropology* 35: 317–335.

Olwig, Kenneth R. 2011. "The Earth Is Not a Globe: Landscape Versus the 'Globalist' Agenda." *Landscape Research* 36 (4): 401–415. DOI: 10.1080/01426397.2011.582940.

Pedersen, Morten. 2009. "At Home Away from Homes: Navigating the Taiga in Northern Mongolia." In *Boundless Worlds: An Anthropological Approach to Movement*, edited by Peter W. Kirby, 135–151. New York: Berghan Books.

———. 2011. *Not Quite Shamans: Spirit Worlds and Political Lives in Northern Mongolia*. Ithaca: Cornell University Press.

Rally for Rangers Foundation. 2017. Webpage accessed January 7, 2018. URL: http://www.rallyforrangers.org. Accessed: Nov. 13, 2019.

Rasiulis, Nicolas. 2016. "Freestyle Bearing: Work, Play, and Synergy in the Practice of Everyday Life Among Mongolian Reindeer Pastoralists." MA diss., University of Ottawa.

Santo, Diana E., and Ruy Blanes. 2014. "Introduction: On the Agency of Intangibles." In *The Social Life of Spirits*, edited by Ruy Blanes, and Diana E. Santo, 1–32. Chicago: The University of Chicago Press.

Stépanoff, Charles. 2012. "Human-Animal 'Joint Commitment' in a Reindeer Herding System." *HAU: Journal of Ethnographic Theory* 2 (2): 287–312.

Strathern, Marilyn. 1992. *After Nature: English Kinship in the Late Twentieth Century*. Cambridge: Cambridge University Press.

———. 2018. "Persons and Partible Persons." In *Schools and Styles of Anthropological Theory,* edited by Matei Candea, 236–246. New York: Routledge.

Tsing, Anna. 2005. Friction: *An Ethnography of Global Connection*. Princeton: Princeton University Press.

U.S. National Parks Service. 2016. "Sister Parks." Webpage accessed September 20, 2016. URL: https://www.nps.gov/yose/learn/management/sisterparks.htm. Accessed: Nov. 13, 2019.

Vainshtein, Sevyan. 1980. *Nomads of South Siberia: The Pastoral Economies of Tuva.* New York: Cambridge University Press.

van Dooren, Thom, and Deborah B. Rose. 2016. "Lively Ethography: Storying Animist Worlds." *Environmental Humanities* 8(1): 77–94.

Wagner, Roy. 1991. "The Fractal Person." In *Big Men and Great Men: Personifications of Power in Melanesia*, edited by Godelier, Maurice, and Marilyn Strathern, 159–173. New York: Cambridge University Press.

Warrack, John. 1980. "Leitmotif." *The New Grove* 10: 644.

Wheeler, W. Alan. 2000. "Lords of the Mongolian Taiga: An Ethnohistory of Dukha Reindeer Herders." MA thesis, Indiana University.

Willerslev, Rane. 2007. *Soul Hunters: Hunting, Animism, and Personhood Among the Siberian Yukaghirs.* Berkeley: University of California Press.

Chapter 8

"I Can't Leave My *Erens*"

Living in a Spirited Geography with Reindeer

Selcen Küçüküstel

Before I first went to the field to do my research among the nomadic reindeer herder/hunter Dukha people in northern Mongolia, one of my biggest challenges was to find their location—not an unexpected surprise when working with nomadic people. I knew that after reaching the closest settlement, I was to enter the taiga, but it was hard to know where the taiga begins or ends, since it has no clear borders. Although most dictionaries describe taiga as a "swampy coniferous forest of high northern latitudes," it was clear that Dukhas did not use the word for forests only. They also used the term to refer to high valleys throughout the region, and for various other corners of the landscape. It seemed to me, they had an imaginary map in their minds; a map outsiders did not have access to. The word taiga was such a significant one, carrying so many meanings with it that from the beginning of my stay, I would hear it coming up in every single conversation. There were songs written for the beloved taiga. All the nighttime tales, which were told in the silence of cold nights, took place in the taiga. Elders, who had settled in the villages, talked about it with a deep sense of longing. Young people, who were curious about other opportunities in the city, happily returned to reunite with their taiga. The taiga was clearly at the center of all discussions relating to life among Dukhas. People never ceased to emphasize that their home "taiga" was perceptibly different from the rest of Mongolia, calling themselves "dwellers of the taiga," and not of the steppes. It was very clear in all regards, that "taiga" referred to much more than a mere geographical area. It comprised of many factors, each of which solicited a strong sense of belonging. Taiga was an area where Dukhas felt at home when they entered—a key part of their political identity, and of who they felt themselves to be.

However, for outsiders, including the government, the taiga that is so critical and full of meaning for Dukhas had almost the opposite connotation. To them it was a "wild" place. The forests of the north, where Dukhas live, spread over a vast geography. The area is usually regarded as "wild and empty" by outsiders, miles away from the nearest urban center. The concept of "wild" in people's minds fits well with this conventional understanding of the taiga. The area is what Lovis would refer to as one of those "big rough spaces with few people," where the population is extremely low, and where mobility is usually very difficult (Lovis 2016, 117–127). This is why the forests of the north are usually chosen for wildlife or survival programs on television channels. Endless forests, wild animals, deep rivers, extreme temperatures dropping to below 40 degrees Celsius in winter, and the difficulty of finding food in the forest, seemed horrifying to most urban dwellers. This image is of course not unique to the taiga in which Dukhas live. The regions in which hunter-gatherers or nomadic peoples reside are often considered empty and wild by outsiders. They are described ethnocentrically, and in ignorance of the history of the place, disregarding the people who have been living there for centuries (Ruotsala 2005, 26).

Dukhas, however, perceive their environment much different from this outsider image. To them, the outsider's "wild geography" is "home" (see Figure 8.1). Yet, for them, home is not wild or dangerous. The concept of home as a "safe" haven, surrounded by walls, is so deeply ingrained in most

Figure 8.1 Autumn Camp of the Dukhas. *Source*: S. Küçüküstel.

non-Indigenous people, that broader concepts of home—as one might witness them among Indigenous nomadic peoples—are difficult to fathom for them. This is why we have endless examples of incidents in which the opinions of people living in those remote areas are hardly taken into consideration, while governments declare them as national parks, or begin huge development projects, thinking—ethnocentrically—that these areas are wild and empty. Unfortunately, the Dukhas experienced a similar problem when their taiga was declared Tengis-Shishged National Park in 2011. At the time, hunting was banned in the area, putting their traditional livelihood and relations with the environment and its animals under serious threat.

This chapter addresses the concept of home for Dukhas; a concept that stretches far beyond the household (or dwelling), and which extends to the whole geography. The chapter looks into the various ways in which the taiga is turned into home. For this purpose, I will first discuss how the landscape is associated with a history of events and of people's movements, which constitute a map of memories. I will then turn to the topic of spirited places. These are locations in the taiga that have their own spirit owners who are often located near ancestral sites in the taiga. I examine these locations in terms of their specific characteristics, focusing on how people regularly interact with these places. Lastly, I explore the spirits who are known to reside in the household (or dwelling), in an attempt of showing how the combination of all these factors turn this vast "wild" geography into a big "home" for Dukhas—a home they share with other nonhuman animals. In the last part of the chapter, the issue of domestication is touched on briefly, particularly in terms of how this attachment to the landscape affects the relations between people and their reindeer.

MAP OF MEMORIES

The word "landscape," probably due to its origins as a technical term used by painters in the late sixteenth century (Hirsch 1995, 2), reminds most people of a beautiful scenic place in nature, where the colors of green and blue shine, ideally with a river nearby, and with mountains in the background. Most of the time, such serene locations are imagined as relaxing and removed from human interaction. Thus, the taiga, where Dukhas live, could easily be considered a beautiful "landscape," relaxing and nice to look at—just like looking at a beautiful painting. Having grown up in a town, this was also how I naturally perceived the taiga: as an incredibly vast and beautiful space, far from human touch.

After beginning to live with Dukhas in the taiga, I gradually realized that all the places that had seemed to me so untouched, were in actuality closely entwined with the human world. When one of the young hunters realized how hard it was for me to describe a place in the taiga in a given conversation, since I had no reference points to talk about places, he decided to help me draw a map of the area. It was at this point that I realized, almost all the mountains, hills, peaks, lakes, rivers, streams, valleys, and taigas had their own names in the Dukha language. These place names were closely related to the stories and past events from which they had derived their names. Learning place names from Dukhas slowly revealed to me some of the knowledge that was embedded in the landscape, and it made clearer to me how people perceived their environment—far beyond the serene landscape I had first envisioned. Since place names are always associated with historical events or stories, it was oral narratives, hidden in place names that had the power to establish strong bonds between people and the natural landscape (Bosso 1996, 40). Thus, the knowledge of place names increased my access to the map of memories in people's minds. Place names also helped me become oriented, as they began to serve me as reference points.

Place names usually have different characters with similar patterns; I will explain them using examples of taiga names. Some place names simply describe the geographical characteristics of the area. An example would be Saylığ taiga, which means "taiga that has a flowing river." Other place names contain practical knowledge about the geography of an area, giving people an idea about what to expect in a place, once they arrive there. An example would be, "Örtten gırhı taiga," which means "taiga with flat burning meadow." People came here in summer, and the area was known for its wildfires. Another example of a place name with practical knowledge was "Gakhba taiga," meaning "taiga with a trap." This area was good for hunting small animals, and people would set many traps there. Some place names related to past events, connecting people with narratives and experiences of their ancestors. For example, "İrool taiga" stood for "man/son taiga." According to legend, a man who had desired a child for a long time, had a son there. So, even today, people who wanted a child prayed in that taiga. Yet other place names had associations with the spirit world, describing the nature of that place. For example, "iili oy" was a small river with a spirit. Similarly, "erenli taiga," was a taiga in which totemic spirit bags had been left behind by the ancestors a long time ago.

As is evident from the examples, the names of the encampments include both geographical and narrative knowledge about the landscape. This enables a young person hearing the name for the first time, to know what to expect in a place, or what the place meant to people. It also enabled access to the collective memory of the society. Naming of geographical locations is one of the most obvious ways for keeping past narratives alive, endowing them with meaning and character. Such places are like the "mental encyclopedias of its users" (Maffi 1999, 28).

In addition to their communal significance, place names also held personal memories for each person living in that taiga. They could evoke entirely different worlds of meaning for anyone. In an effort to better understand the kinds of associations place names brought into people's minds, I began asking Dukhas about their favorite places in the taiga, and why they liked them. My inquiries brought out so many detailed memories and stories attached to those places that, after a while, even the most ordinary geographical features came to life. The land had become to me like a precious book, holding vital knowledge about the people. As each individual had a different mental map in their mind, their surroundings became imbued with unique meanings. It was obvious that the same mountains or forests, that to an outsider might have seemed wild, for the Dukha observer constituted a completely different environment—a place filled with memories from every valley, hill, river, or forest in which they had spent a lifetime. Some people might have been born in one of those valleys, others had first fallen in love near a certain stream, or given birth in a certain valley. Yet others would have formed a new household for the first time in a particular taiga. The role of hunting trips and nomadic movements was evident in the formation of this sentimental relationship with the landscape. I remember how forty-two-year-old Çokhu, who was a very quiet person, got really emotional when he was talking about his favorite taiga, which he had visited with his father and uncle when he was seventeen years old. They had stayed there for almost ten days. Çokhu had hunted his first deer there, and the trip was one of the best memories he had of his father before he passed away a couple of years later.

I listened to so many stories like this during the time I stayed in the taiga that the landscape almost came to serve as a visual research tool. I would simply point to a geographical feature on our trips through the taiga and hear a story or a memory about it. Moreover, the memories about the landscape were not all about people. There were also many narratives about occasions where someone had met a spirit, or where something significant had happened for a reindeer or a dog. Many incidents involve other-than-human animals or spirits. Thus the landscape did not only incorporate the human world, and memories relating to it. As a whole, the taiga ecosystem involved people, spirits, and nonhuman animals. With those life experiences, ordinary trees or rivers turned to stories and past experiences. In other words, the geography was "understood as events, rather than discrete geographical locations" (Aporta 2016, 69). Since Dukhas do not have written sources or history books to talk about their past, all their information is embedded in the landscape. Turning every corner in a valley, or climbing a small hill by a river, all these memories were refreshed again and again, and knowledge was re-remembered.

In addition to all those memories, another factor making the landscape both historically and sentimentally an important place for Dukhas was their burial custom. Dukhas traditionally practice sky burials, leaving their dead in the open, allowing animals to eat the corpse. Burying the dead was considered a taboo, and when asked, most people would say that digging up the ground was a form of pollution. Some people explained this in terms of the pollution caused by the decaying body in the soil. While bones remained in the ground when buried, they disappeared in sky burials. Others explained the tradition in terms of reciprocity with the animals. Since people eat animals when they are alive, animals should eat people when they are dead. Animals, especially birds, such as eagles, ravens, or vultures, consumed the dead. It was important to people that the ritual was conducted appropriately, or else the soul of the person might not leave, in which case it could seek to disturb the living. The funeral rite is named *curttadir* in Dukha, which literally translates as "to make someone settle." The ceremony transformed the geography into an open cemetery, or as the name in Dukhan suggests: it turns it into a locality in which the beloved ancestors "settle." Thus, people wandering around this vast landscape know that it hosts the spirits of the deceased, and they make offerings when passing by places where they know the deceased have been left.

Considering all these factors, it is easy to understand the role of the landscape for one's life, as there are many incidents embedded in the geography. People have a map of memories in their mind, a tender and intangible map that far exceeds geographical features. This is why when Dukhas look at the taiga, they do not simply see the land, but they see a combination of everything, ranging from precious moments spent there, to spirited places, and even an open cemetery, where they had to leave their loved ones.

SPIRITS ARE EVERYWHERE

Beyond evoking memories and past events, features in the landscape were also considered alive for Dukhas. As is the case for many other animistic societies, this meant that the geography they lived in was filled with spirits. Living in the taiga consequently required more that survival skills. It called for a detailed knowledge about the nature of these spirits and how to interact with them. David G. Anderson explains this relationship, using the concept of "sentient ecology," in which hunters act and move in full expectation that animals and tundra will respond to their actions (Anderson 2000). Dukhas similarly attributed the power to react and interact to their surroundings. As there are many types of spirits that reside in the taiga, proper interaction requires knowledge of their specific character. From this perspective, all entities in nature have powers, being part of nature (Humphrey, Mongush and Telengid 1993, 53). When people walked around in forests, valleys, lakes, or rivers, they would communicate with other beings all the time, or at least acknowledge their presence. In order to survive, people had to maintain good relations with these powers (lngold 2000, 66–67).

The main universal spirit among Dukhas is called *cer-eezi*, which literally means "land or world owner," the supreme power and mediator between all living things. It is the cer-eezi that protects the animals or people in the taiga and punishes them when they violate the proper rules of conduct among themselves. Cer-eezi is considered to be the omniscient spirit controlling the land, but people believe that certain places have their own spirit owners, and one has to be careful in those places so as not to offend them. People have to interact with these "intentional non-human agents" in a friendly and submissive manner if they wish to avoid trouble (Pedersen 2009, 138). Described as "sacred places" by some researchers, these locations are usually "connected with what the western world classes as 'natural' features of the 'landscape', such as mountain peaks, springs, rivers, woods and caves" (Carmichael, Hubert and Reeves 1994, 1).

Dukhas have different words to describe spirited places and their different characters. Again, the aim is for someone to know what to expect in a place. Having heard its name, they are able to act accordingly. Such places were usually described as "soft" *cımcak* or "hard" *gatı*. Soft places could be visited, but they required special care. People had to make sure offerings were made to the spirits residing there. Hard places, on the other hand, could not be visited, but people would make offerings to them from far away. Different opinions existed as to why spirits in specific places had become malevolent. Some stories claimed that those places had been polluted by people a long time ago, and thus the spirit of the place had become aggressive. Other stories claimed that these places had always been like this, because these were areas that needed protection. The character of a place could also be "tied to human interaction with the land and its spirits" (Kristensen 2004, 34). Areas where hard spirits lived could be normal places at ordinary times, but when angered, they possessed the power to change the weather or intervene in other ways. For example, if the spirited place was a river, and somebody stepped into it, it might flood or quickly turn into a raging torrent. If the place was a mountain and someone attempted to climb it, the wind might pick up dramatically. One of the best-known mountains in the region, Mount Arig, was famous, among Dukhas, for its malevolent spirit. A well-known story told of a team of researchers who had attempted to climb the mountain. Soon after they had set out, a snowstorm swept across the mountain. The researchers were forced to return to camp, leaving one of their team members behind. The Russian researcher who was left behind did eventually make it back to the camp. On his return he was pale, looking visibly terrified, and despite his gallant effort in making it back to camp, he died soon after. Nobody had attempted to climb the mountain since. To my Dukha friends, this incident proved that anyone who tried to climb there would eventually come to suffer.

Such stories were not just confined to the past. Many Dukhas have had similar experiences for themselves. For example, there was a lake behind one of the Dukha encampments, and the spirits around that lake were known to be very angry. People always asked for permission from the spirits before they drank the water. It was considered extremely dangerous to waste water in such a place. People were to take only enough water to drink, but never to transport water away from its source. According to Dukha tradition, there were also spirited places that were not open to hunting, as the animals living here were under the protection of cer-eezi. The consequences of hunting in a forbidden spirited place could be devastating, and many stories attested to hunters who had suffered miserably from breaking these taboos. A middle-aged Dukha hunter explained to me how he had suffered for many years after hunting an elk near a spirited river he had not known about. On his way back from the hunt, six of his reindeer died in a wolf attack, while he lost many more animals from his herd in the days that followed. For many years following the incident, he had suffered from a recurring nightmare featuring an elk that was attacking him. Other recollections of broken taboos in spirited places told of punishments, including the death of a hunter in the sacred place itself. Everybody seemed to agree that hunting in spirited places was akin to stealing from another person—an act that could have lasting consequences, not only for the perpetrator, but also for generations to come.

On the other hand, soft spirited places had a calmer character, and people could visit them, making offerings, and even benefitting from their healing power. One of these places in the taiga was famous not only among Dukhas but also among Mongolians. When I first arrived in the taiga in the summer of 2012, I met an old man who was traveling with his daughter to Arıkshan, a spirited river that is famous for its ability to heal various diseases. Despite its remote location, many people visited Arıkshan in summer. Another similar place with a soft character was Coşama Lake, where Dukhas sometimes camped in autumn. People spoke of a small island in this lake, and from time to time, a white horse would appear on it. Seeing this white horse brought good luck to the person, and it meant protection for them and their family.

Another well-known soft spirited place was Caraş River, which was known to produce stones in different shapes. There were many stones in the shape of animals in the river, and keeping one of those stones could bring good luck to the person, or it could even protect them. However, this softness should never be interpreted as an excuse to ignore breaking a taboo such as hunting. Soft places were usually associated with good luck and fortune, but they always had to be treated with great respect, much like one would expect a hard place, since any type of spirited place had the potential to harm or provide fortune to people.

Apart from soft and hard places, which were tied to spirits in various ways, there also were places referred to as *azalı cer*. This term literally translates as "haunted or evil place." Such places were not necessarily tied to a specific landscape, but they had the capacity of changing over time, which made them even more treacherous. An area usually became haunted, because someone had passed away there, and their soul had never left the place. They had usually died there as the result of neglecting proper spiritual conduct. As those areas were not stable, people needed to be able to recognize the signs of a haunted place. Strange incidents could occur in haunted places. For example, someone might accidentally hurt themselves, or a dog might bark without reason. People also experienced unusual nightmares, or they could hear unusual sounds in such locations. Hard places were easier to avoid, because people learned about them from their elders, but haunted places could be anywhere, and people understood where they were only after they had come to experience it.

PROTECTORS OF THE LAND: AVILAN AND BATIKSHAN

Although all spirited places protect their area from humans and harmful activities, there are some legends among Dukhas that mention special spirits, who reveal themselves to people, as they become visible or tangible. Two of these famous spirits are Avilan and Batıkshan. These two spirits are of the forest and of the mountains, and they protect the animals that reside among them, primarily from hunters trying to kill in spirited sites. Especially Batıkshan, who appears in the form of a short woman with long hair, was well known among Dukhas, and there were many stories about how she had revealed herself to people. According to the most famous legend about Batıkshan, a man had gone hunting when he came across a deer. He wounded the deer, but the deer did not die, and it kept running away toward Mount Arig. He followed the deer to the mountain to kill it, at which moment the weather suddenly turned. A big storm broke out, and it started to rain and snow. The man realized he had entered a spirited mountain place, and filled with fear, he returned home. On his way back, he saw the deity with his own eyes. The spirit was a short woman with long hair, and she was holding half of a bowl in her hand. She was healing the deer that he had wounded, and it seemed to recover. Scared by the incident, the man ran away (Kücüküstel 2013, 98).

In most of these accounts, Batıkshan is healing animals or protecting her mountain or a forested area from people, and even from domesticated reindeer. People have seen her chasing domestic reindeer from an area, or scaring a hunter who is about to kill on a spirited site. People have also seen her while sitting on a fallen tree in the forest, or while drinking water from a river. Elders stated she did not reveal herself to everybody, especially these days. In the stories, people were usually afraid when they saw her; however meeting Batıkshan was also interpreted as a good omen. One woman, Enkhtuyaa, told me that many years ago, her father had seen Batıksshan while wandering around the taiga. He had seen a little child sitting on a tree, a child with long fuzzy hair. He rushed back to the camp to tell the people about it. The elderly people said that he had seen a land spirit, and that it would serve him well. When he decided to go back to invite the spirit to the camp, so they could host her, she had already gone. People told him that this was a sign for him being lucky and that he would have a long life.

Yet another land spirit that appeared to people, if not as frequently as Batıkshan, is Avilan. Like Batıkshan, this spirit protected the environment, but according to legend, it had a dual form, resembling a half human and half animal. As the account goes, all people have a good and a bad side, and if this spirit comes across a person, it will divide that person into two, taking the good side for itself. The footsteps of this spirit are human on one side and bear on the other, representing the duality of Avilan (Kücüküstel 2013, 99–100). Pedersen, who also discusses Avilan in his work on Dukhas, talks of other legends where a hunter who killed a wild reindeer near Mount Arig annoys the mountain spirit. The hunter feels very scared but has to spend the night there since it is getting dark. The next day, when he wakes up, he sees several half people climbing the mountain, and the dead reindeer he had killed the previous night getting up from where it had lain. The Avilan heals the animal, returning life to it (Pedersen 2009, 139).

While most Dukhas were fearful of Avilan and Batıkshan, as they could be dangerous if one had broken a taboo, many also respected them. The core of this respect was related to their sense of justice, because people knew that these spirits were a threat only when people did something wrong. Their sometimes aggressive ways of guarding their territory was to be perceived as venerable, even necessary.

DAHILGA: ANCESTRAL SITES

The discussion so far has focused on spirited places in the taiga, places that make no discrepancy between people's ancestral lines. However, there do exist places that are specific to Dukha, depending on their family lineage, which are called *dahilga*. These ancestral sites were usually places where a powerful shaman ancestor from that family was laid to rest, using the sky burial. It could also be a specific tree, a mountain, or a valley that was considered an ancestral site. The children of a family would mark their respect each year by making offerings in this place, and if this was not possible, then they would make offerings in direction of the dahilga. Just like spirited places with an owner, there were two types of dahilga among Dukhas: the soft kind and the hard. If one's dahilga was hard, family members could not physically visit the ancestral place because the spirit would become upset if someone were to approach. In this case people just made their offerings from afar. They would set up an *örgıl*, a special worship totem to show their respect, and make offerings.

Families who belonged to a soft dahilga could visit their sites. Here they would, pray, tie khadag, and make offerings. The spirits in those places did not only like to be visited, they expected being visited. Knowledge about one's dahilga was passed on from one's ancestors, and whether soft or hard, hunting was forbidden near all ancestral sites. During my fieldwork, not all people could visit their dahilga, even if it was soft, as some of the sites were located on the other side of the border in Tyva. People tried to compensate for this in various ways, but it remained an issue of great importance to some elders. Gantulga, one of the elders in the east taiga told me how important this was to him:

> Dahilga is of great importance for the wellbeing of a family. Our ancestral spirits dwell there, and praying for them regularly is extremely important. For example, my dahilga is a large shaman tree in Tyva. We used to go there every year with my father and make our offerings. It was a four-day journey on reindeer, but every autumn it was a tradition for my family. After the border between Mongolia and Russia was closed and it became strictly controlled, we could no longer visit our tree. We try to pray from here but who knows what we are missing by not going there. Sometimes, when we have bad fortune, I can't help but think it might be because of that. (Gantulga, 04.06.2015)

In addition to certain ancestral sites, most families among Dukhas also had a dahilga tree, which belonged to their family lineage. Visiting a dahilga tree was necessary for the men, as they had to make offerings there regularly. The rituals of visiting a dahilga tree are similar to visiting any dahilga, and the best time to visit one's sacred tree is in the autumn, within fifteen days of the new moon. Once the day of the journey was chosen, people went to the tree, approaching it carefully, following the movements of the sun, moving clockwise. They would make their offerings and offer their petitions. Most people claimed that within eighteen or twenty-one days they could tangibly feel a change in their lives. According to most elders, the changes that occurred under socialism had broken the links between families and their trees, causing adverse effects on those families who had experienced dislocation between them and their tree. Sickness among families, and especially in children, the elders claimed, had increased since this segregation. Many people attributed special value to their tree and thought that their tree affected their lives in many ways. They claimed that if they were not to visit their tree for a prolonged time, the consequences would become very obvious; things would start to go wrong. An old hunter told me that around ten years earlier, he had realized that his hunting success was directly related to his family tree, suggesting his neglect had been reflected in his lack of hunting success. While the dahilga brought good luck to people when they visited them, they also had the potential to harm if they felt neglected for a long time.

I CAN'T LEAVE MY ERENS

It is evident from these examples that, like many Indigenous peoples around the world, Dukhas "extend the concept of sacredness to the whole of their land" (Hubert 1994, 16). Similarly, the concept of home is also in a way extended to the whole geography. Changing homes quite often, setting their home up, and taking it down several times every season, Dukhas learn to become attached to the landscape, rather than the "home" itself. This attachment gives them a sense of belonging to the land. The distinction between "home" and "outside" is not as sharp as with city dwellers. However, this does not mean that they do not feel relieved when they see the smoke of tents or barking dogs of a camping site after a long trip in the taiga. Individual dwellings are also of great importance in many aspects. Although life in the taiga is very hard—far from the comforts of a settled house—people who go to the village even for a short time always return, complaining about how much they missed their home in the taiga. I believe, all the aspects of the geography I discussed above make clear why people are so attached to this landscape. Everything, including their past, their memories, ancestors, and spiritual connections, is embedded in this landscape.

When I spent a long winter among the Dukhas in 2015, on a cold evening, sitting by the fire with Sarnai, I found out how this connection was beyond my perception; etched into the hearts and minds of the people. It was about to get dark, and we were preparing food. Sarnai had been on the phone with her daughter for more than ten minutes when the connection cut out and she was complaining about the instability of the mobile phone network in the taiga. Sarnai, who was fifty-six and had three daughters, had been living in the taiga almost all her life. Despite having children, she never got married. One of her daughters lived in the village with her husband and children, and Sarnai talked to them at least a few times a day when the network was available. They talked about the details of their day: Sarnai would tell her daughter about the reindeer. Her daughter, Tunga, would tell Sarnai about her grandchildren and what they were doing at school. Since they were so fond of each other, I asked Sarnai that evening why she did not move to the village for the winters, staying with them, as some people did. Since the reindeer were kept in a faraway valley in the winter, there seemed to be no reason for her not to move to the comfort of the village. To my surprise, she said she could not go because she could not leave her *alazy* (Dukh. for *dwelling*). Unable to understand what she meant, I looked at her in puzzlement. Then she pointed to the backside of the tent, and observing that I was still confused, she said, "The *erens*. I cannot leave because of the *erens*. I cannot leave them."

An *eren* in Tyvan is a protector spirit, a kind of shamanic totem. Most Dukhas had spirits in the form of erens at their home, and there was a special place in the household for erens, usually opposite the door (see figure 8.2). Physically, the eren is a piece of fabric that has been animated by a shaman, to turn it into a protector spirit. When people had a serious problem, they would go to the shaman, and after performing a ritual and communicating with the spirits, the shaman would disclose the source of the problem. Depending on the nature of the problem, sometimes at the end of the ceremony, the shaman fashioned an eren for the family or for a specific person, to protect them. Some erens were intended for a specific person, others could be for the whole family, and some were made for a reindeer. If a person was having problems individually, for example, if a small child was suffering from an illness, the shaman could make a special eren for her or him. Similarly, if a family was experiencing difficulties with their herd, a shaman could tell them to select a sacred reindeer to protect the rest of the herd, or the family (cf. chapter 6, this volume). In this case, the eren was made for the animal. Sometimes erens could be made specifically for a certain aim, like a hunting eren, which would bring luck during the hunt. While most erens resided in spirit bags in the household, the eren made for a sacred reindeer was tied to the animal in the form of a collar.

Once an eren had been formed, the person needed to take care of it all their life, renewing the eren with a ceremony every spring. The ceremony necessary for the eren would change, depending on the situation and family lineage. Some families could make their eren only from dark colors; even the thread they used had to be dark, and they updated their eren only at night when the stars had come out. They would use only black tea without milk for their offerings. There also were families who had to use only white fabric. The ceremony took place early in the morning at sunrise. On rare occasions, a shaman could make an eren for someone but tell them to keep it in the household on behalf of another relative, such as a parent or grandparent. In either

Figure 8.2 Erens in a Tent of the Dukha Camp. *Source*: S. Küçüküstel.

case, the eren became almost like a family member, living with the person for the rest of their life. The erens in the household, along with the spirit of the hearth, were considered the owners of the dwelling. This is why people made offerings to erens and to the hearth constantly, offering milk tea in the mornings. When a gift was brought home, including game meat, the first piece was offered to the hearth and the erens, and only after that did people start to eat. Hunters always prayed to the erens at home before departing on a hunt.

The number of erens in a household varied, and some families, especially younger ones, sometimes had none. According to tradition, when a couple got married among Dukhas, [Duk. *öglenir*, literally to "get a tent or home," an *ög*], they formed a new household, which always started out without erens. The beginning of a new household was represented by building a new hearth in this household together; forming a new unit. Because the spirit of the fire was considered the owner of the household, it was very important that the couple build their own hearth, giving life to the household. Along with a new fire, a couple would have their own erens in the household, since none could be brought from their parents' household. In this sense, every family created their own history through erens. Erens thus served as family history, or album, indicating the main struggles of the family, and marking the times when they needed protection.

The reason why Sarnai was so sensitive about her erens was that she had never gotten married. Without a husband, she had never formed a new dwelling of her own. Instead, she was keeping her parents' household spirits, maintaining the same household with its erens and hearth. In total, she had twelve erens; two of them were for the sacred reindeer, one belonged to her elder sister, another to her niece, she had one for herself, there was a hunting eren from her family, and the others were for her children and grandchildren. Since people left their erens—even their personal ones—in the household of their parents after they got married, Sarnai, being single, was keeping the erens of both her sister and brothers, as well as those of her own children. In this way, her home tied together multiple generations. All of these people would visit her tent from time to time to pray and make offerings to their protector spirits. Since she was also holding the *olca eren* (the hunting eren) of her family lineage, all the men in her family, among them being her brother and nephews, came to her household to pray before they went on hunting trips. Following every successful hunt, the men would hang a part of the animal on the eren. This could be a piece of hair, hooves, or claws. By keeping alive the fire of her parent's household, including their spirits, Sarnai was serving as a medium connecting the past to the present. This was why she could not leave her household to go and live in the village. Even if her move were to be temporary only, she would be leaving behind a huge responsibility. I had never realized, up until this point, how weighty a responsibility Sarnai had been carrying, and why it had been so difficult for her to leave the taiga. Leaving the taiga, for her and for most other elders, meant abandoning who they were. It meant leaving behind their spiritual connection, and their identity, breaking the family chain and its history.

LANDSCAPE AS A COMMON GROUND CONNECTING PEOPLE AND REINDEER

Having discussed the factors that connect Dukhas to their landscape, I would like to conclude this chapter by illuminating some of the ways in which this intimate relationship with the landscape is also closely related to the concept of domestication. Beyond doubt, the relationship between people and their domesticated reindeer is deeply affected by the relationship they have with the land, almost to the point of forming the backbone to domestication. Let me explain.

Traditionally, Dukha subsistence had mostly been focused on hunting as a means of subsistence, and reindeer had played a supplementary role, being slaughtered for meat only in times of scarcity. As such, Dukhas were practicing the Sayan type of reindeer keeping (Vainshtein 1980, 130). The main functions of the domesticated reindeer were transportation, riding, and milk production. Most stories and myths about the origin of reindeer among

Dukhas support these facts, mentioning how Dukhas were living in the taiga as hunters and when they met the domesticated reindeer, it made their lives much easier, because people could now survive in the taiga more easily by riding the reindeer and migrating to far-away places. All of these stories have a similar theme in which reindeer help people who experience difficulties surviving in the taiga (Küçüküstel 2013).

The foundation of this relationship, in which reindeer assist people in surviving in the taiga, remains very true to this day. Reindeer are the only domesticated species that can survive in the taiga, thanks to their ability to move in deep snow and in freezing temperatures, foraging for themselves (Vitebsky 2005, 20, 23). Indeed, it is the only domesticated animal that is native to the taiga, and people are dependent on its cooperation if they wish to maintain the taiga as their home. In other words, the fact that both parties, reindeer and people, perceive the taiga as a common home is what brings them together.

When Dukhas say, "We cannot live without the reindeer," they mean it literally. Life in the taiga requires reindeer, as it is impossible to migrate without them. Much like Sarnai, who cannot leave her erens if she wants to remain in the taiga, so Dukha also cannot leave their reindeer if they wish to stay in their homeland. Of course, people can move to the village and herd other livestock, much like other Mongolian families do. But it is their deep connection to the taiga that, in turn, makes people depend on their reindeer. Thus, I believe, it is this deep, yet often neglected, connection that plays such a crucial role in the continuation of reindeer breeding.

The whole issue of domestication is too complex to discuss in detail here. But, in short, the fact that people need reindeer more than reindeer need people—providing that the people want to live in the taiga—changes the nature of relations between people and deer in terms of how domestication is practiced. Most theories of domestication focus on how herders apply dominance and control to their animals, as opposed to hunters who may rely on an element of trust (Ingold 1986; Ullah 2005). More recent theories of reindeer domestication mention terms such as symbiotic domestication, in which reciprocity between people and animals is considered a core feature, as both parties learn from each other and affect each other's behavior (Beach and Stammler, 2006). However, all these theories focus on the practical sides of daily interaction between people and reindeer, while failing to look at the big picture.

I will argue that the control Dukha exert over their reindeer is a nurturing kind of control. People may seem to hold power over their reindeer in limiting the freedom of the animal by tethering it from time to time, and by focusing on ways to protect it from dangers such as wolf attacks. However, the reindeer hold a sentimental kind of power over people in this relationship, as they maintain their status as natives of the taiga who help people stay in their beloved geography. In other words, the landscape is the core feature holding reindeer and people together, rather than common spheres of either tameness or wildness that might serve as a shared "home" for both parties. The commitment of Dukhas to their reindeer is fundamentally connected with their attachment to their landscape. This is why I suggest that, when evaluating the relationships between people and animals, landscape and vicinity are of great significance, offering another dimension to anthropological debates on human-animal relations, apart from functional, economic, or social significance.

CONCLUSION

As we have seen from all the examples above, the taiga is considered alive for Dukhas. People have always interacted with their environment and with the spirits residing in all corners of the land. I believe, the idea that people are always being watched when they are in the taiga, brings along a unique existential perception for Dukhas, and perhaps for other animistic societies. I would like to explain this with a quote from John Berger's famous book *Ways of Seeing*, interpreting his idea in a different way. Berger writes: "Soon after we can see, we are also aware that we can be seen. The eye of the other combined with our own eye to make it fully credible that we are part of the visible world. The reciprocal nature of vision is more fundamental than that of dialogue" (Berger 1972, 9).

In this paragraph, Berger is likely speaking of other people, when he writes "the eye of the other." However, it supplies a perfect example for how Dukhas perceive the world, except that this "eye of the other" is not always another human being for them. I agree with Berger that the "eye of the other" is something that makes us feel visible and even alive, as our existence is constantly acknowledged in this way. But in Western societies, this eye of the other only consists of other people, which means that our interaction with nature misses the "other eye," which predisposes us to people much more than to places in the environment. For a society like the Dukhas, the environment consists of many "other eyes," so a person always feels validated and visible while walking in the taiga. Mountains and rivers constantly interact with them, seeing them in much the same way as they see mountains or rivers.

This naturally turns the action of seeing into a reciprocal act, and it increases the attachment to the environment, turning it to an active, mutual process in which diverse beings always acknowledge each other. The fact that one is always seen and visible to the spirits of the taiga while wandering around, results in people never feeling lonely in the taiga, even if there are no other humans around them. The attachment to the landscape is very strong, and when a Dukha person leaves the taiga, they are really shaken to the core, their visibility and existence threatened, as they become less visible in life.

REFERENCES

Anderson, David G. 2000. *Identity and Ecology in Arctic Siberia: The Number One Reindeer Brigade*. Oxford: Oxford University Press.

Aporta, Claudio. 2016. "Markers in Space and Time. Reflections on the Nature of Place Names as Events in the Inuit Approach to the Territory." In *Marking the Land, Hunter-Gatherer Creation of Meaning in their Environment*, edited by William Lovis and Robert Whallon, 67–89. Oxon and New York: Routledge.

Beach, Hugh, and Florian Stammler. 2006. "Human-Animal Relations in Pastoralism." *Nomadic Peoples* 10(2): 6–30.

Berger, John. 1972. *Ways of Seeing*. Based on the BBC television series with John Berger. London: Penguin Books.

Bosso, Keith. H. 1996. *Wisdom Sits in Places. Landscape and Language Among the Western Apache*. Albuquerque: University of New Mexico Press.

Carmichael, David, Jane Hubert, and Brian Reeves. 1994. "Introduction." In *Sacred Sites, Sacred Places*, edited by David L. Carmichael, Jane Hubert, Brian Reeves, and Audhild Schanche, 1–8. London: Routledge.

Hirsch, Eric. 1995. "Landscape: Between People and Space." In *The Anthropology of Landscape. Perspectives of Place and Space*, edited by Eric Hirsch, and Michael O'Hanlon. Oxford: Clarandon Press.

Hubert, Jane. 1994. "Sacred Beliefs and Beliefs of Sacredness." In *Sacred Sites, Sacred Places*, edited by David L. Carmichael, Jane Hubert, Brian Reeves, and Audhild Schanche, 9–19. London: Routledge.

Humphrey, Caroline, Marina Mongush, and B. Telengid. 1993. "Attitudes to Nature in Mongolia and Tuva: A Preliminary Report." *Nomadic Peoples*, 33, 51–61.

Ingold, Tim. 1986. *The Appropriation of Nature: Essays on Human Ecology and Social Relations*. Manchester: Manchester University Press.

———. 2000. *The Perception of the Environment: Essays on Livelyhood, Dwelling and Skill*. London: Routledge.

Kristensen, Benedikte Moller. 2004. "The Living Landscape of Knowledge, an Analysis of Shamanism among the Duha Tuvinians of Northern Mongolia." Master's thesis, Institut for Antropologi, Københavns Universitet.

Kücüküstel, Selcen. 2013. "The Dukha and How They Perceive the Environment: Maintaining Good Relations with Landspirits." Master's thesis. Yeditepe University.

Lovis, William. 2016. "Network Maintenance in Big Rough Spaces with Few People. The Labrador Innu-Naskapi or Montagnais." In *Marking the Land, Hunter-Gatherer Creation of Meaning in their Environment*, edited by William Lovis and Robert Whallon. London: Routledge.

Maffi, Luisa. 1999. "Language and the Environment." In *Cultural and Spiritual Values of Biodiversity. A Complementary Contribution to the Global Diversity Assesment*, edited by Darrel Addison, 22–36. London: Intermediate Technology Publications. UNEP.

Pedersen, Morten Axen. 2009. "At Home Away from Homes: Navigating the Taiga in Northern Mongolia." In *Boundless Worlds: An Anthropological Approach to Movement*, edited by Peter Wynn Kirby, 135–152. Oxford: Berghahn Books.
Ruotsala, Helena. 2005. "In the Reindeer Forest and on the Tundra. Modern Reindeer Management and the Meaning of Local Ecological Knowledge." In *Culture and Environments*, edited by Anepaio Terje, Runnel Pille and Vallikivi Laur. *Pro Ethnologica* 18: 23–48.
Ullah, Isaac. 2005. "Human-Animal Interactions and the Shift from Hunting to Herding." Master's thesis. University of Toronto.
Vainshtein, Sevyan. 1980. *Nomads of South Siberia: The Pastoral Economies of Tuva*. Cambridge: Cambridge University Press.
Vitebsky, Piers. 2005. *Reindeer People: Living with Animals and Spirits in Siberia*. Great Britain: Harper Collins Publishers.

Chapter 9

Killed by Mutton

On the Powers of Animal Products in Dukha Households

Benedikte Møller Kristensen

During spring 2010, the atmosphere in the camp of a group of Dukha reindeer herders of Mongolia was tense. This was particularly the case in the households of three related Dukha reindeer herding families. Though they seldom voiced their anxieties in public, worry was written all over their faces, and they appeared wary each time one of their kin left the camp. One evening Aichurek, an elderly Dukha woman, confided in me, telling me the reason for her and her kin's worries. She told me how deaths had haunted her own household and those of her siblings over the past three spring seasons. None of them had died within the camp, but all during travels within and beyond the taiga. A woman and a man had died of apparently natural causes (a heart attack and a stroke), while the bodies of three young men were discovered with bruises indicating they had suffered violent deaths. One of these men was Aichurek's own son, who, as many other Dukha men in those days, had traveled to Baruun taiga in search of gold. Shortly after his arrival at the mining site, his brother in law found him dead in a mine with a broken skull.

With a quiet voice Aichurek confided, that she thought the series of deaths haunting the households of her family was triggered by a piece of red raw mutton meat, which her cousin, the shaman Mönhöö, had improperly offered to an eren (a talisman enlivened with spirits), to which all of these households were tied through their fellow kinship. In the camp, several rumors circulated about how this or that family member's improper dealings with various nonhuman agencies, both within and beyond the household, had triggered this series of deaths. Most of these rumors concerned the visit of the shaman Mönhöö, who although of Dukha descent had never lived in the taiga. A Darhad family from the steppe had adopted Mönhöö as a child, and she had lived most of her adulthood in Ulaanbaatar. However, in 2008 she had travelled to visit the shaman Ariunaa, the younger sister of Aichurek, to pay homage to the eren, which her deceased father, the shaman Mergen, had enlivened approximately sixty years earlier to protect the lives of Aichurek and her siblings. Ariunaa kept this eren in her yurt, as she had never married and was said to continue the hearth of her parents, including the eren of their children. Ariunaa often told me how she lamented that she had not stopped Mönhöö from feeding her eren with meat. She whispered, "After this, my own son died. They found his dead body in the lake in Tsagaannuur."

The question is why Aichurek and Ariunaa, as well as other Dukhas, felt that a piece of meat offered to an eren could trigger this series of human deaths. In order to answer this question we need to examine how Dukhas perceive the very handling of animal products within the domestic sphere of the human household as instituting and shaping a network of relations between spirits and humans within and beyond the immediate human household.

Aichurek told me how she had been in her sister Ariunaa's yurt the day Mönhöö came to pay homage to the eren. She recounted:

> She [Mönhöö] told us she had become a shaman and then she placed plates of meat and a bottle of vodka in front of our eren. It looked very disrespectful. My father never offered meat to our eren. "It is prohibited to offer it meat"; he used to say. We were never even allowed to handle meat in the northern area of the yurts [hoimor]; it was always handled far from the eren. For some families, it may be fine to make offerings of meat, as nothing bad will happen, but our family is not like this. But we did not say anything, because she was a shaman, so we thought maybe she knew better. [. . .] If meat is placed close to the eren it turns ferocious and [. . .] it will not recognize the human worshipper as kin, but see her as a stranger. That is why Mönhöö died shortly after. For some families nothing bad will happen if they offer meat, but our family is not like this, because our eren is an eren with an eagle claw and it is "allergic" to [Mon. harshildah] meat.

The story of Aichurek hints at the erens of different households or kin-units. Due to their unique origins, specific human and animal spirits had dissimilar traditions for what constituted their proper diets and appropriate ways of handling animal products within their vicinity. This appears to be a classic example of Claude Lévi-Strauss' (1963) famous theory on totemism, namely how Indigenous peoples engage natural species to represent kinship units, not because they "are good to eat," but because they are "good to think," as species that can be engaged as classificatory forms to denote divisions within human society. Among Dukhas, divisions between kin-units, such as families, lineages, or clans, were indeed understood through their origin with different erens (See Fugure 9.1), whose unique kind of being and power could commonly be

Figure 9.1 Elder Dukha Woman Showing her Inherited *Erens*. *Source*: B. M. Kristensen.

traced back to an origin that lay with a specific shaman and a particular wild animal. Yet, Dukha eren do not simply configure as devices to think through. Rather, I propose that we view the eren from a multispecies lens and explore how eren in the shamanistic ontology of Dukhas configure as devices for the mutual becoming of humans and nonhumans. As Donna Haraway (2003) has suggested, animals are not merely "good to think," but also "good to live with," as human becoming inevitably is a becoming with animals.

This is evident in the way Dukhas perceive the very character of their eren as being continually shaped by the way its human kin interact with it, that is, by what the eren is fed and how animal products are handled in its vicinity. Likewise, Dukhas believe that the nature of the eren shapes the lives of its human kin, which was what Aichurek was referring to when she proposed that the anger of her eren had triggered a series of deaths among her kin. Among Dukhas, the position as kin of a certain eren is not pre-ordained, but it is a multispecies becoming, which involves the risk of becoming a stranger to one's own eren. Not only the act of feeding an eren, but also the mere handling of animal and other products within the yurt is said to shape the character of the eren and determine whether the eren will view the worshipper(s) as kin or as strangers. Although the relation between eren and humans is enacted within the domestic sphere of the yurt, Dukhas believe that its affect extends into the wild forest, where it determines whether its sentient beings will view the human wanderer as kin or as stranger.

Whereas many multispecies ethnographies have revealed the mutual benefits and creative outcomes of multispecies encounters (Haraway 2003; Kirksey & Helmreich 2010), studies on shamanism (Fausto 1999; Pedersen 2013) have shown how Indigenous peoples from around the world perceive multispecies encounters as vital for life, yet also as dangerous, as they entail the risk of becoming other. I now return to Aichurek, who told me that the eren of her kin was "allergic" to raw meat, and that it had turned into a *chötgör* [a demon] upon being fed meat, and subsequently had started to haunt its former kin as they moved within and beyond the taiga. She further told me how the eren, although being kept in the household of her younger sister, was linked to all the sacrificial places of their kin in the wild forest, and how the spirits of those places, who used to protect their kin, had now turned against them. The challenge of Dukhas appears to lie in the fact that they must interact with nonhumans, such as animal products, in a way that somehow confirms the "mutual being" (Sahlins 2011) of a specific human kin group and certain nonhumans, while simultaneously affirming the "mutual otherness" (Stasch 2009) of this specific human kin group and other nonhumans.

In this chapter, I explore why the handling of various animal products within the human household of the yurt is thought by Dukhas to animate the very being of humans and nonhumans who belong to the household of the spiritual entity of an eren. I show how Dukhas perceive the feeding of the eren, as well as the mere management of animal products within the yurt, as inherently risky, because it is through these acts that the eren recognizes its human kin either as true kin or as strangers. I argue that the animistic ontology of Dukhas is directed by a principle of what I term inter-animation, referring to how nonhuman entities, such as animal products, when brought into the vicinity of an eren of a specific household are thought to facilitate a series of animations of nonhuman and human beings, within, between, and beyond the human household(s). I argue that the Dukha yurt constitutes what David Anderson et al. have termed an "architecture of relationships" (2017) and, what I term an "architecture of division," designed both to (re)produce the mutual being of human and nonhuman kin, and the mutual otherness of human and nonhuman non-kin.

LIVING IN A SENTIENT LAND

Today there are around 500 Dukhas living in the remote Tsagaannuur district in the northwestern corner of Hövsgöl Province of Mongolia. Being a Tyvan minority, their mother tongue is "Dukhan," an "endangered" Tyvan dialect of

Turkish origin (Ragagnin 2006). However, today the majority of those who are able to speak Dukhan are over the age of sixty; those below this age tend to speak Mongolian, although they understand Dukhan. At the time of my fieldwork in 2012, only around 200 Dukhas, approximately forty households, were living as reindeer herders and hunters in the areas of Baruun Taiga (the western taiga) and Zuun Taiga (the eastern taiga), north of Tsagaannuur village. The remaining 300 Dukhas had either adopted a more settled lifestyle in the district center of Tsagaannuur, or they herded steppe-based livestock in the steppes bordering the taiga.

The traditional livelihood of Dukhas was characterized by a distinctive combination of taiga reindeer husbandry and hunter-gathering (Vainshtein 1980 [1972]; Johnsen et al. 2012). Meat obtained from wild game, and milk received from semi-domesticated reindeer, historically constituted the main diet for Dukhas, who supplemented this with wild vegetables, herbs, and berries. Hunting and herding has historically been part and parcel of the shamanic tradition of Dukhas, as wild game and healthy herds were thought to be the "blessing" [*hishig*] of *Oron Hangai* [the taiga] and its spirits, only bestowed upon those, hunters and herders, who interacted in spiritually appropriate ways with the land and its spirits.

During socialism many Dukhas were forced to resettle in the villages of Tsagaan Nuur and Ulaan Uul where they worked in the local *negdels* [Mon. collectives] as fishermen or as lumber workers. Only a limited number of Dukhas stayed in the taiga, where the local negdels employed them as state herders or hunters. However, the Democratic Revolution in 1990 led to the closure of the negdels in Tsagaannuur and Ulaan Uul and left Dukhas without work. Subsequently, many Dukhas, some of whom had lived in the villages for thirty years, decided to leave their settled life and moved back to the taiga, where they resumed their former livelihood as reindeer herders and hunters.

Among Dukhas, as among many Indigenous peoples in Siberia and Mongolia, the end of socialism also marked a return to their shamanic tradition. Scholars have tended to view shamanism after socialism through the lens of post-socialism, as an effort to reconstruct identity and meaning, which were felt to have been lost, following the decline of socialism (Buyandelgeryin 2007: 142; Buyandelger 2013; Lindquist 2005; Shinamura 2004), or as "an ontology of transition" actualizing the labile and uncertain character of post-socialist society (Pedersen 2011: 35). In contrast Dukhas turn toward shamanism not primarily as an Indigenous reaction to the political transition, but as a consequence of Dukhas returning to the landscape (the taiga) and practices (hunting and herding) in which their spirits subsist.

Dukhas are divided into four exogamous patrilineal clans (*udam*)—Balygsh, Soyan, Urat, and Zoot—of which the Soyan and Balygsh clans are further subdivided into several exogamous patrilineal sub-clans (also named udam). Each clan is divided into several shamanic lineages [Mon. *böögiin udam*] and smaller kin-units of families, each of which have their own talismans (erens) that are kept in the household and in sacrificial places, located in the wild forest. These are enlivened with specific human and nonhuman spirits, commonly a historic shaman, and a specific wild animal. Each kin-unit has certain traditions [Mon. *yos*] for proper interaction with the land and its sentient beings. The content of these traditions is constituted by the nature of the kin's eren, which, however, often is uncertain, whereby acts of traversing the land and interacting with its beings, and even the handling of animal products, is felt by Dukhas as circumscribed by risk.

Dukhas commonly depict their return to the taiga as a return to Oron Hangai; literally the forested hilly land, which involves shamanic ideas about the land being a "sentient ecology" (Anderson 2000). As such, it watches and acts upon those who traverse it. When Dukhas return to the taiga, they relocate to the cosmic economy of Oron Hangai. Those whom Oron Hangai recognizes as kin are blessed with the gifts of prey and healthy herds, whereas those recognized as strangers may be hit with misfortune. It is through interacting with the land and its beings, according to the innate traditions of its places and beings, as well as according to those of the herder or hunter's kin, that Oron Hangai comes to see the human as either a fellow kin or a stranger. In my former work (Kristensen 2015) I have focused on how Dukha shamanism is enacted as if it were in the wild, namely through the practices of herding and hunting. In contrast, this chapter will move into the domestic sphere of Dukha yurts, to explore how the handling of animal products within the yurt also contributes to establish or break kinship bonds between human and nonhuman households within and beyond the domestic sphere.

ARCHITECTURES OF RELATIONSHIP AND DIVISION

Scholars have long studied the organization of domestic space in Mongolia. They have shown how Mongolians order and engage domestic space to represent and (re)enact their social organization, gender divisions, kinship, and local ideas of the cosmos (Empson 2007; Humphrey 1974; Sneath 2000: 216–221). Turning to Dukhas, the spatial organization of the yurt resembles that of the Mongolian *ger*. The northern, or rear section of the yurt is known as *hoimor,* and it is treated as the most honorific and purest part of the yurt. The southern sphere, the section where the entrance is located, is seen as the least honorific and most impure area of the yurt. The eastern side of the yurt is used for cooking, and it is a place usually confined to the household, whereas the western part is used for cleaning, and is the place where guests are seated. The stove is placed in the middle of the yurt. It hosts the domestic hearth, which is linked to the lives and fortunes of the household. It is forbidden to burn garbage in the hearth, which must be kept pure at all times, as it is regularly fed the upper and most honorary part [*deesh*] of meals and of milk-tea. From a wooden pole in the northeastern part of the yurt, right where the domestic area meets the honorific area, hang the talisman(s) of the household's eren(s). They are either hidden away in a cotton bag, or fully on display.

Elder guests and family members are usually seated in the honorific north, while younger people are seated in the less honorific south. The Dukha yurt, much as the Mongolian *ger*, constitutes a domestic place that re-enacts social and kinship positions in the human realm. Nonetheless, I argue that the yurt also constitutes an architecture of relationships and divisions that extend into the nonhuman realm, as the Dukha erens link the domestic place of the human household to the nonhuman households of the wild forest, and as the act of handling animal objects within the yurt both serves to connect and disconnect humans and nonhumans.

Rebecca Empson (2007) has shown how wooden chests in the Mongolian household constitute a network that facilitates the continuation of human relations, which could not otherwise be enacted in a shared space. She argues that in Mongolia it is not so much people who constitute a home, as the nomadic livelihood of Mongolians entails a continual movement away from home, and between homes. Instead, parts of kin members' bodies, such as pieces of umbilical cords and children's hair from the first hair-cutting ceremony, are kept in the household chest, where they constitute a site through which relations to absent kin are perpetuated and integrated within the home.

In the case of Dukhas, I argue that, although the eren is placed in a particular human household, it constitutes a network of relations between all its human kin who, although they may not all live in the same household, are tied to the same eren due to their former or present worship of it. Additionally, the eren is tied to one or more sacrificial places within the wild forest. These are places that have their roots in one or more ancestral shamans. As the worshipper feeds and prays to his lineage eren, the Dukha says that his prayers are passed on to his sacrificial places. This ensures that upon approaching them, he will be recognized as kin. The act of worshipping and feeding the eren then works to perpetuate ties to sacrificial places that are located in the wild forest, often far away from the camp. Through this process, spirits of the wild forest become integrated in the home, while humans become integrated in the wild forest.

However, the nature of the relation between eren and humans, and thus between sacrificial places and humans, is not given beforehand. Instead it is shaped through human acts of feeding the eren and through the way animal products, as well as other objects, are handled within its vicinity. The story of Aichurek reveals how a simple piece of meat offered to her kin's eren was thought to trigger a series of deaths. The Dukhas often say that the land of Oron Hangai protects them, as they traverse its lands, because it perceives them as kin. However, the human position of kin to the land of Oron Hangai is fragile, as the case of Aichurek's siblings reveals. Where kin members make the mistake of offering an inappropriate gift, they and their relatives may fall into the position of a stranger to the land of Oron Hangai. The act of making an offering to an eren, which somehow does not fit with its inner nature, is to establish a dangerous relation. Instead, the handling of animal products that may not agree with the nature of the eren of the household ought to be handled in the southwestern, or sometimes just the southern part of the yurt. These areas are places for less honorific guests, and for non-members of the family. By using the yurt's architecture to help separate certain entities from others, the members of the household demonstrate to the eren who its kin are. They visually acknowledge that certain animal products or social connections are unbecoming to the eren. Here we are dealing with what Morten Axel Pedersen (2013) has termed a "labour of division aimed not at doing away with relations, but as a technique of doing something with relations" (2013: 202).

Most Dukha households share certain rules for how animal products ought to be handled within yurts, to harness a beneficial relation between human and nonhuman households. For instance, most households commonly fry meat from wild game directly on the iron top of the hearth, whereas they abstain from doing so with meat from steppe livestock (sheep, goats, cattle, horses). I was told, that wild game meat is pure and clean, because wild game feed on wild herbs and plants of the taiga, and because the wild animals and Dukhas share a fellow origin in the taiga. In contrast, meat from livestock is considered potentially impure and contrary to Dukhas. This is because the character of their human owner may still linger in the meat, and because Dukhas do not share a common origin with the livestock that comes from the steppe. Cow, mutton, goat, and horse meat all ought to be treated with caution, and they were not to be placed directly on the iron top of the hearth, nor could they be placed in the hoimor. Where this did happen, it could arouse the anger of the spirit. This was what had happened with Aichurek's kin, where several people had died during their travels back and forth between the taiga, as the land of Oron Hangai had come to see them as strangers or as others.

The act of handling animal products within the yurt not merely shapes the content of the relation between humans and the nonhuman household(s) of Oron Hangai, but also the relation between humans and the household of their reindeer. This became evident when Dukhas warned me not to mix reindeer milk with meat during the spring. Spring was the period when reindeer had their young, and when their young were suckling milk. In winter, Dukhas would frequently heat up a leftover meal of meat by pouring hot milk tea over it. This was an unthinkable act in spring, as the mixing of milk and meat was believed to reduce lactation in reindeer. It could also induce illness in the herd, because milk and meat were thought to be allergic to each other, given their diverse origins. Although none of my Dukha informants could quite account for how this mixture of substances within the yurt could trigger an effect in the milk production of reindeer, everyone was convinced of its causality, which many had experienced.

This was also the case of Otgon, a Dukha woman in her fifties, who I one day overheard scolding her two young grandchildren for pouring hot milk tea into a leftover meal of *tsuvian* [fried noodles] to warm it up. I could not hear what exactly she was saying, but from the tone of her voice and from the faces of the children, I could see that they clearly had done something wrong. The next morning, I noticed that Otgon was examining one of her reindeer calves, which had just died. I later asked her why the calf had died and she answered:

> Yesterday Nomgon and Enhbayar came around and took some of my *tsuvian* and poured milk tea over it. And this morning the bowels of my calf swelled up and then it died. [. . .] Things like this happen if you break the rules of your clan.

The words of Otgon illustrate how the Dukha consumption of reindeer milk commonly is arranged according to the dietary rules of the eren attached to certain clans. It is an example of how cases of sudden illness or death among the herd often are believed to be caused by people having ignorantly broken various rules for handling animal products.

The very physical location of yurts within the camp also reflects an architecture of relation and division, as it both is aimed at (re)establishing relations between members of the same kin-groups or clans and at avoiding relations between members of dissimilar kin-groups or clans. During the start of my fieldwork, I often wondered why one of the fifteen households at the camp would always raise its yurt at a distance of around 500 meters from the remainder of the households. This was the household of Otgon and her three daughters. I went to visit her one day to find out why she placed her yurt like this. As I arrived at her place, I could see her tethering her reindeer calves to wooden poles next to her yurt. Having tied the last calf, Otgon and I went into her yurt. For a while we talked about the beauty of the land surrounding the campsite. Then I asked Otgon, why she had raised her yurt so far from the rest of the camp. She answered:

> In my Soyod clan we always raise our yurts far away from those of others, to prevent people from walking through the fold or behind the hoimor. If someone

> walks through these places with a saddled reindeer, or with a packed reindeer, our livestock will [not] grow. But people differ from each other. In some families nothing happens if someone passes the hoimor or the fold with a saddled reindeer. In our family it is also prohibited to step over the poles used for tying up the reindeer during the fowling season. Not even the owner of the reindeer is permitted to step over the poles. The people of the Soyod clan have such customs [*zanshil*] and this is why we do not walk across other people's hoimor or folds.

That Otgon had placed her yurt at a far distance from the rest of the camp was thus not a matter of coincidence. Instead, this was how she established a kin relation with the eren of her clan, requiring her to engage in a labor of division that would safeguard her herd from potential harm, which Otgon knew could befall her clan if people were to walk across a members' hoimor or fold. The rest of the camp did not appear to be concerned with such risks. The brother of Otgon, the shaman Batzaya, had even placed his yurts right in the middle of the camp and appeared not to view this as a threat to his reindeer. Later the wife of Batzaya, Bolormaa, explained to me that she was caring for her reindeer calves according to the traditions of her family (clan/lineage), rather than those of her husband's family. This was, according to Bolormaa, because nurturing reindeer calves and milking reindeer cows generally followed the traditions of the housewife's family, as these tasks were usually undertaken by women.

WHEN FEEDING IS BECOMING KIN OR STRANGER

Much like the Mongolian *ongon* (Hangartner 2011: 71; Purev & Purvee 2005; Pedersen 2007; Vainshtein 1978: 458) the Dukha concept of eren refers to a certain human and nonhuman spirit helper, and its materialization in a particular talisman, sacrificial tree, or shamanic costume. Yet, they also differ, as ongons primarily are of human origin (Pedersen 2007, 2011), whereas erens tend to be both of a human and nonhuman (animal or natural element) origin. According to Sevyan Vainshtein (1978: 458), Tyvans traditionally differentiated between anthropomorphic and zoomorphic erens. During my fieldwork, I did not find such differentiation. Instead, my informants commonly traced the origin of their erens back to a certain animal (wolf, wolverine, bear, lizard, etc.), or to a natural element (I only heard of water) whose powers were thought to fuse with those of its ancestral shamans. Yet, the overall form of the eren is similar to that of the ongon, which consists of a "multiplicity" (Even in Pedersen 2007: 155) of spirits encapsulated in a single form. Dukhas' eren may, drawing on Pedersen's (2007) definition of the Darhad's ongon, be defined as:

> [A] complex amalgamation of events, social relationships and material substances which continuously absorb ever more events, social relationships and material substances in accordance with the particular spatio-temporal trajectory defined by the original event from which the spirit was created. (Pedersen 2007: 155)

This definition reveals how the character of the ongon, as the eren, is labile, yet somehow bounded to its origin in a certain spirit. Among Dukhas, each eren is thought to have the particular nature constituting the totality of its nonhuman and human spirits. For instance, an eren of eagle origin is thought to have eagle-like powers, enmeshed with the specific powers associated with the historical disputes, misdeeds, or heroic deeds of its ancestral shamans or other influential ancestors. Once a kin member has fed an eren of his or her kin, a binding relationship of "entrustment" [*daatgax*] to the eren is formed. This relationship has to be perpetuated through regular renewal [*shinechilj*] of the eren, which can be accomplished by tying cotton ribbons to it, and by feeding it with the food that is in agreement with its innate nature. As Janet Carsten (1995) has argued, "personhood, relatedness and feeding are intimately connected" (1995: 224), as it is through feeding that strangers become kin. In our case, it is through feeding that the relation between humans and nonhumans is constituted as one of "mutual being." However, feeding may also trigger the opposite. It can have a dividing effect, making those who once were kin into strangers. This is what happens when an improper animal product comes into the proximity of an eren. At the same time, for Dukhas to be recognized as kin of their eren requires a certain "labour of division," which ensures that particular animal products are handled at a safe distance from the eren. This labour asserts the "mutual otherness" of the eren from the animal product that may be of a somewhat contrary nature to it.

Most Dukha households have one or more erens, some of which are connected to a particular individual within the household, whereas others may relate to several or all members of the household. There are other erens, which are tied to the members of several households. These erens are connected to the siblings and offspring of those who shared a household at the time the eren was first enlivened. This was the case of Aichurek's eren, which was made by the shaman Mergen, when Aichurek had been just a toddler. The eren had been intended to protect Aichurek and her siblings against all kinds of misfortune. Although, at the time of my fieldwork, the eren was kept in the household of the shaman Ariunaa, the sister of Aichurek, it continued to affect the lives of Aichurek and her siblings, as we have seen.

Ariunaa told me that she only fed white and pure foods to the eren. White foods, such as milk tea, dairy products, and incense from wild juniper were fit for feeding it, as the power of the shaman Mergen and of the eagle were both white and pure. I learned that food offered to an eren was to resemble that of its origin, as other kinds of food could provoke anger or fear in the eren. A male elder by the name of Bataar once told me how he always fed his eren with black tea. The eren had its origin with a crow, and as he said, "crows drink black water." Although food offered to the eren ought somehow to resemble its nature, it must not be of the exact same nature, as this too would provoke fear in the eren. For instance, Otgon told me how the eren of her kin had originated with a bear. Otgon could not quite explain why, but her eren also possessed white power, which had to be fed with reindeer milk or milk tea. When I asked her whether she could feed it with bear meat, Otgon looked horrified and answered, "No, no. It is a bear eren, so it would be terrified if it were presented with bear meat. It would even get frightened if bear meat were to be brought into the yurt." To offer a bear eren a piece of bear meat is thus thought to imply a cannibalistic act, which is likely to frighten and possibly anger the bear eren.

I learned from Otgon, that the bear eren protected her and her kin not only within the domestic sphere, but also beyond it. She recounted how her elder brother had once been suspected of theft. On his way to the steppe, he had heard that the police were after him to imprison him. While resting in the taiga, he fell asleep. In his dream, he was walking down a taiga trail to reach the steppe. However, he soon discovered that a sleeping bear was blocking the way. When he had found another trail to avoid the bear, he soon was confronted by another bear, blocking the new trail. Later, he heard that the police had turned around their car, and that for some reason they had decided not to charge him with the suspected theft. Upon completing the story, Otgon said, "You see, our eren 'protects' [*hamgaaldag*] us from anything. It watches over us when we travel in the land of Oron Hangai." The story shows how the act of feeding an eren enlivens the "mutual being" and its protective relationship with kin. It also tells us about the responsibilities of feeding it with animal products of a similar nature to that of its own, while withholding animal products of a nature too similar or too dissimilar to those of its origin.

However, the challenge of Dukhas is that they often do not know the nature of their eren and are uncertain about how to appropriately interact with it. The anger of the eren could be provoked if something of an alien nature, including a human being belonging to kin unrelated to that of the eren, is brought into its vicinity. For instance, only kin members are ordinarily allowed to touch or sleep next to their eren. They are also the only ones permitted to make offerings. However, because it is the erens who make decisions about who is kin, Dukhas can never be certain of their heirship, even when among their own. For instance, one of my elderly informants, Bat, told me how he had placed the eren of his deceased father in the wild forest. He had felt that it was unsuitable for him to keep, as it had been made for his father. Shortly after, Bat experienced all kinds of misfortunes. First he broke his leg, then his son fell severely ill, and finally wolves killed several of his reindeer. Confused by these incidents, and anxious about whether it had been wrong to give up his father's eren, Bat visited a shaman. The shaman told Bat that his hardships were the result of having wrongly abandoned his father's eren, which had been passed on to him. Bat then retrieved the eren from the place where he had left it and brought it back to his home in order to resume his ties to it.

ALLERGIC SPIRITS

During my fieldwork, I was puzzled by Aichurek's statement about her kin's eren being allergic to meat. When I asked her to elaborate on this, she simply replied, "Mutton meat is a bad thing, because it raises the *chötgör,* haunting our kin." She then repeated, "Our kin are allergic to meat." My puzzlement on the concept of allergy further grew when others repeatedly invoked the same notion to explain why it was prohibited to mix various animal products, or why certain animal products could only be handled in the southern sphere of their yurts.

The Dukha concept of *harshildah* translates as "being allergic to," or "being contrary to." I have chosen to use it in the allergic sense, as Dukhas use the term to denote how certain animal products cause detrimental reactions, including the transformation of character, in other beings (humans and nonhumans). As such, the term shares similarities with the more conventional term of allergy by referring to reactions occurring when a person's "immune system reacts to normally harmless substances in the environment" (Shin et al. 2013: 22). When Dukhas talk about their erens being allergic to certain animal products, they are obviously not referring to a reaction within the immune system of the eren. Rather, I found that what they are saying is that certain animal products much like 'allergens' may be harmless or suitable to certain erens, while causing detrimental reactions in others, depending on the nature and history of the given eren. We are thus dealing with an animistic ontology, which appears to be directed by a principle of what I term "inter-animation." This refers to how animal products may animate the nature of an eren, which in turn is thought to animate the very lives of humans.

A common Dukha term for an animal product that is thought to cause allergic reactions is "bad thing" [*muu yum*]. Some animal products are perceived as being bad for anyone if they somehow are polluted by human misdeeds. Other animal products may become good or bad, or simply neutral, depending on how and by whom they are handled. Scholars have described how Mongolians perceive wild prey obtained through hunting as either potentially pure or polluted, depending on whether the hunter interacted properly, or improperly, with the land and its spirits during the hunt (Empson 2011; High

2008). In this example we are dealing with a kind of deferred or "secondary human agency" (Gell 1998: 20), in which the intentions, emotions, or acts of human beholders bestow products of wild prey with agency. Although Dukhas share such perceptions, what is important to note here is that "bad things," which are thought to elicit "allergic" reactions, are not considered bad for everyone everywhere. Instead, such bad things are relational. Their status arises from encounters between the innate nature of an animal product and an eren.

The piece of raw mutton meat, which shaman Mönhöö placed in front of Ariunaa's eren, was not a dangerous agent in itself. On the contrary, it became dangerous when Mönhöö placed it in front of Ariunaa's eren as an offering, because Ariunaa's eren, having an eagle origin, was allergic to mutton meat. Hence, the case shows how animal products, such as raw mutton meat, become dangerous (or "bad things") when they encounter an eren of a contrary nature. People told me that all Dukha erens were allergic to the meat of livestock from the steppe, because of their origin in the taiga, as well as their association with wild animal species and ancestral shamans (Dukha). By contrast, Darhad ongons required offerings of mutton, as their origins lay with sheep and cattle, and with ancestral Darhad shamans. This may also be why Mönhöö, having been raised by steppe nomads, found it suitable to feed her ancestral eren with mutton.

The above reveals how the Dukha notion of "bad thing" is not an isolated event. Rather, it is reminiscent of Mary Douglas' (1995 [1966], 36) classic definition of dirt as a kind of "matter out of place." Douglas defined dirt as a "by-product of systematic ordering and classification of matter" (Douglas 1995 [1966], 36). While the "bad thing" accentuates the differences between natures of people (Darhad and Dukha) and kin-units, it does not merely represent social and spiritual differences. On the contrary, a "bad thing" has the capacity to transform the nature and character of erens, and hence it reveals animistic principles rather than purely classificatory ones. The "bad thing" may thus be understood in accordance with Gell's (1998) definition of an "agent" whose character "is a function of the social-relational matrix in which it is embedded" (1998: 7). Contrary to Gell, I do not perceive the agency of the "bad thing" as entirely a matter of deferred or abducted human agency, but as a kind of animated byproduct of its encounter with the innate nature of an eren.

Dukhas, however, face the challenge of never being quite sure of the innate nature of their eren. It is only through living in the taiga, and by experiencing its protection and blessing or its delivery of misfortunes, that people obtain insight into whether the handling of specific animal products within the domestic sphere has turned them into kin or strangers to the land of Oron Hangai and its spirits.

CONCLUSION

This chapter has shown how the eren in the domestic space of the Dukha household constitutes a network of relations that connect to places beyond the human household. It connects the households of semi-domesticated reindeer with those of spirits located in specific places within the wild forest. I have argued that Dukha yurts constitute an architecture of relationship and division between various kinds of beings. The very handling of animal products is guided by shamanic principles of "mutual being" and "mutual otherness" of certain humans and specific nonhumans. Beings or products of a similar, but not too similar, nature ought to be placed in the same area of the yurt. Beings or products of a dissimilar nature ought not to be mixed, but must be kept in separate areas of the yurt. Among Dukhas, multispecies encounters are vital for becoming kin, as it is through feeding the eren with appropriate animal products that humans are recognized as kin, not only by their eren, but also in the eyes of Oron Hangai. Yet, multispecies encounters are also circumscribed by danger and risk, as the simple act of bringing an animal product into one's household may elicit the anger of one's eren, which can turn one into a stranger of the taiga.

REFERENCES

Anderson, David G. 2000. *Identity and Ecology in Arctic Siberia: The Number One Reindeer Brigade*. Oxford University Press.

Anderson, David G., Jan Peter Laurens Loovers, Sara Asu Schroer, and Robert P. Wishart. 2017. "Architectures of Domestication: On Emplacing Human-Animal Relations in the North." *Journal of the Royal Anthropological Institute* 23 (2): 398–416. doi:10.1111/1467-9655.12613_1.

Buyandelger, Manduhai. 2007. "Dealing with Uncertainty: Shamans, Marginal Capitalism and the Remaking of History." *American Ethnologist* no. 34(1): 127–147.

Buyandelger, Manduhai. 2013. *Tragic Spirits: Shamanism, Gender and Memory in Contemporary Mongolia*. Chicago: University of Chicago Press.

Carsten, Janet. 1995. "The Substance of Kinship and the Heat of the Hearth: Feeding, Personhood, and Relatedness among Malays in Pulau Langkawi." *American Ethnologist* no. 22(2): 223–241.

Douglas, Mary. 1995 [1966]. *Purity and Danger: An Analysis of Concepts of Pollution and Taboo*. New York: Routledge.

Empson, Rebecca. 2007. "Separating and Containing People and Things in Mongolia." In *Thinking Through Things: Theorising Artefacts Ethnographically*, edited by A. Henare, M. Holbraad, and S. Wastell, 113–140. London & New York : Routledge.

———. 2011. *Harnessing Fortune: Personhood, Memory and Place in Mongolia.* Oxford: The British Academy: Oxford University Press.

Fausto, Carlos. 1999. "A Blend of Blood and Tobacco: Shamans and Jaguars Among the Parakanã of Eastern Amazonia." In *Darkness and Secrecy: The Anthropology of Assault Sorcery and Witchcraft in Amazonia*, edited by N. L. Whitehead and R. Wright, 157–178. Durham: Duke University Press.

Gell, Alfred. 1998. *Art and Agency: An Anthropological Theory*. Oxford: Oxford University Press.

Hangartner, Judith. 2011. *The Constitution and Contestation of Darhad Shamans' Power in Contemporary Mongolia*. Leiden: Global Oriental.

Haraway, Donna J. 2003. *The Companion Species Manifesto: Dogs, People, and Significant Otherness*. Chicago: Prickly Paradigm Press.

High, Mette. 2008. "Wealth and Envy in the Mongolian Gold Mines." *Cambridge Anthropology* no. 27(3): 1–18.

Humphrey, Caroline. 1974. Inside a Mongolian Tent. *New Society* 30(630): 273–275.

Johnsen, K. I., B. Alfthan, P. Tsogsaikhan, and S. D. Mathiesen. 2012. Changing Taiga: Challenges to Mongolian Reindeer Husbandry. *Portraits of Transition* NO 1, United Nations.

Kirksey, S. E., and Helmreich, S. 2010. The Emergence of Multispecies Ethnography. *Cultural Anthropology* 25(4): 545–576.

Kristensen, Benedikte Møller. 2015. "Returning to the Forest: Shamanism, Landscape and History among the Duha of Northern Mongolia." PhD Diss., University of Copenhagen.

Lévi-Strauss, Claude. 1963. *Totemism*. Boston: Beacon.

Lindquist, Galina. 2005. "Healers, Leaders and Entrepreneurs: Shamanic Revival in Southern Siberia." *Culture and Religion* 6(2): 263–285.

Pedersen, Morten A. 2007. "Talismans of Thought: Shamanist Ontologies and Extended Cognition in Northern Mongolia." In *Thinking Through Things: Theorising Artefact Ethnographically*, edited by A. Henare, M. Holbraad, and S. Wastell, 141–167. London: Rouledge.

———. 2011. *Not Quite Shamans: Spirit Worlds and Political Lives in Northern Mongolia*. Ithaca, NY: Cornell University Press.

———. 2013. The Fetish of Connectivity. In *Objects and Materials: A Routledge Companion* edited by Penny Harvey, Eleanor Conlin Casella, Gillian Evans, Hannah Knox, Christine McLean, Elizabeth B. Silva, Nicholas Thoburn, Kath Woodward, 197–207. London: Routledge.

Purev, Otgony, and Purvee Gurbadaryn. 2005. *Mongolian Shamanism*. Ulaan Baatar.

Ragagnin, Elisabetta. 2006. "The Position of Dukhan Among the Tuvan Dialects." In *Exploring the Eastern Frontiers of Turkic*, edited by Marcel Erdal and Irina Nevskaya, 153–156. Wiesbaden: Harrasowitz Verlag.

Sahlins, Marshall. 2011. "What Kinship Is (Part One)." *Journal of the Royal Anthropological Institute* 17(1): 2–19.

Shinamura, Ippei. 2004. "The Movement for Reconstructing Identity through Shamanism: A Case Study of the Aga-Buryats in Mongolia." *Inner Asia* 6(2): 197–214.

Shin, Hee Soon, Min-Jung Bae, Sun Young Jung, and Dong-Hwa Shon. 2013. "Inhibitory Effect of Skullcap (Scutellaria Baicalensis) Extract on Ovalbumin Permeation in Vitro and in Vivo." *Food Chemistry* 140 (1–2): 22–30. doi:10.1016/j.foodchem.2013.01.042.

Sneath, David. 2000. *Changing Inner Mongolia: Pastoral Mongolian Society and the Chinese State*. Oxford: Oxford University Press.
Stasch, Rupert. 2009. *Society of Others: Kinship and Mourning in a West Papuan Place*. Berkeley: University of California Press.
Vainshtein, Sevyan I. 1978. "The Erens in Tuva Shamanism." In *Shamanism in Siberia*, edited by V. Diózegi and M. Hoppal, 457–468. Budapest: Akadémia Kiadó.
———. 1980 [1972]. *Nomads of South Siberia: The Pastoral Economies of Tuva*. Cambridge: Cambridge University Press.

Part IV

SOIOTS, OKINSKII DISTRICT, REPUBLIC OF BURIATIA

Chapter 10

Jade Hunters

Local Connections to Land and Resources in Okinskii District of Buriatia

Anna Varfolomeeva

The beginning of my fieldwork in Okinskii District of Buriatia[1] did not go according to plan. I intended to focus my research on local relations with mining development, but it seemed that nobody was willing to talk to me about gold or jade extraction carried out in the district. The local residents kept saying that all the extraction sites were situated far away from the settlements, and they did not know much about the activities of extraction companies. Only after being in Oka[2] for some time, did I realize that the seeming absence of the resource extraction theme was more of a façade. In fact, many of the locals in Oka are involved in the extraction of jade, but this extraction is conducted informally and kept a secret. This realization made me look at the familiar settings of the settlements differently, and soon I understood that jade extraction was in fact present and visible—though not openly.

This chapter focuses on a case study of informal jade extraction in Okinskii District of Buriatia, south central Siberia. This remote region is rich in mineral resources, and mining exploration there dates back to the nineteenth century. Since the early 2000s, informal jade extraction has become widespread in Okinskii District. The extracted stone is transferred to Irkutsk or Ulan-Ude and sold to Chinese buyers. While jade has no historical value in Buriatia, in China it is in high demand, and this difference in valuation has influenced the growth of informal mining activities. While jade is not highly valued in Oka beyond its economic importance, the close connections of local residents with the land have an impact on their perceptions of jade extraction. As various parts of the landscape, including underground, are animated and connected to spirit masters, informal extraction of jade becomes not simply a business enterprise, but a spiritual journey. This chapter analyzes the complex perceptions of jade in Okinskii District as well as the parallels between the recent phenomenon of jade extraction and more traditional local economic activities such as hunting.

As Tim Ingold (2007) notes, the specific properties of materials are not their attributes, but their histories, as materials get their qualities from direct interactions with the physical world. Wet pebbles look and sound differently from dry ones, Ingold points out, as their "stoniness" changes as they interact with the environment (Ingold 2007, 15). Materials may also interact with the spiritual landscape of a certain territory, as Nils Bubandt (2017) demonstrates in his study of entanglements between stones and spirits in Indonesia. This chapter focuses on the interrelations between jade extracted in Buriatia and its surroundings, both physical and symbolic. It demonstrates that jade mining does not disrupt the close connections of Oka residents with their landscape, but instead has become an integral part of these connections.

The chapter is partly based on a subsection of my PhD dissertation focusing on the articulation of Indigenous subjectivities under the influence of mining discourses in two Russian regions: Karelia and Buriatia (Varfolomeeva 2019). The chapter relies on the data of participant observation and semi-structured interviews in Orlik, Sorok, and Khuzhir settlements of Oka and in Ulan-Ude, the capital of Buriatia. The fieldwork was conducted in two phases, in the summers of 2016 and 2017. During the fieldwork in Buriatia, I conducted interviews with local residents, representatives of the local administration, as well as with activists and NGO representatives. My initial contacts were my hosts at the local guesthouse in Orlik, the main settlement of Oka, as well as the representatives of the local administration in Orlik. Later, I used a snowball sampling method to enlarge the number of potential consultants. In total, I conducted thirty-six interviews in Buriatia, but this chapter is primarily based on the interviews conducted with the residents of Oka who are directly involved in informal jade extraction.[3] To secure the informants' anonymity, I use pseudonyms when referring to interview participants throughout the chapter. I have also changed some of the informants' data to make sure that they are not identifiable within the community. In addition to interview materials and informal conversations with local residents during my fieldwork, the chapter is based on publications, including local and regional newspapers such as *Akha*,[4] *Inform Polis*, and *Novosti Buriatii*.

In the following sections of the chapter, I will discuss the sacred relations of the residents of Oka with their landscape. I will then focus briefly on the history of mining development in the district and discuss the notion of resourcescape, which I use as a unifying term connecting different types of natural resources in the landscape, while blurring the boundaries between them. The final section of the chapter demonstrates that jade is perceived by locals as a subsistence resource similar to hunting, and therefore the "modern" activity of jade extraction has become embedded in the pre-existing human-landscape relations in Okinskii district. Through informal jade extraction, the residents of Oka manage various types of negotiations with the spirit masters of the territory, but also with state authorities who try to control the jade extraction business in the area.

SYNCHRONIZATION WITH THE LANDSCAPE

Oka is situated in the western part of Buriatia and borders Irkutskaia Oblast' (to the north), Mongolia (in its southeastern part), Tyva Republic (Russia), and Tunkinskii District of Buriatia. The name of Okinskii District comes from the Oka River, a tributary to the Angara River. The river passes through most of Oka's settlements, including Bokson, Sorok, Orlik, Saiany, and Khuzhir. While the district is the fourth largest in Buriatia, based on its territory (26,594 square kilometers), it is very scarcely populated and has only 5,400 residents. The mountain chains surrounding Oka make travels to the district especially difficult (Pavlinskaia 2002, see Figure 10.1).

Oka is one of the most remote parts of Buriatia. By car it takes around twelve hours of travel from Orlik to Ulan-Ude, and approximately eight hours from Orlik to Irkutsk. Although Oka is closer to Irkutsk than to Ulan-Ude, traveling to Irkutsk is in many cases more difficult. There is a daily minibus [Rus: *marshrutka*] connecting Ulan-Ude and Orlik, but no organized transportation to or from Irkutsk. Due to remoteness and transportation difficulties,

Figure 10.1 The Mountains Surrounding Orlik. *Source*: 2016, A. Varfolomeeva.

Oka is often perceived by those outside the district as a special, "sacred" place, a land full of legends, or as the land of wilderness and untouched nature. The climate of Okinskii district is relatively harsh. In winter, the temperature may go below minus 40 degrees Celsius, and snow may stay from November until May. The warm summer period rarely exceeds forty days, and even during this time, nightly frosts may occur (Pavlinskaia 2002). As summers are not always warm, it is difficult to grow vegetables. As a result, local residents of Oka rarely engage in agriculture.

The population of Okinskii district is mostly made up of Oka Buriats and Soiots. These two ethnic groups, while residing side by side, have culturally influenced each other (Pavlinskaia 2002). In 2000, Soiots received the official status of Indigenous Small-Numbered People of the Russian Federation. Linguistically, Soiots are close to Tofas and Dukhas (Mongush 2012). However, it is almost impossible to draw a clear dividing line between Buriats and Soiots in Oka. As Vakhtin, Golovko, and Schweitzer (2004, 256) note in their study of Russian Old-Settlers in Siberia, "the contemporary ethnic and cultural reality in the Russian North and in the whole Arctic region is undoubtedly an intermixture. The binary systems of classification, still widespread today (indigenous-non-indigenous) are in the best case a sincere fallacy." This chapter views Oka residents as a unified community of local dwellers with blurred boundaries between different ethnic groups.

The relations between humans and landscapes in Oka are embedded in a syncretic Buddhist-shamanist belief system (Belyaeva-Sachuk 2012). Buddhism and shamanism play vital roles and are in many cases perceived interchangeably. Thus, before embarking on a difficult trip, Oka residents may turn to a shaman or lama in order "to open the road," for example, to ask the spirit masters of the territory to permit the trip. Eduardo Viveiros de Castro views shamanism as a crossing of ontological borders "to adopt the perspectives of nonhuman subjectivities in order to administer the relations between humans and nonhumans" (Viveiros de Castro 2004, 468). Similarly to this definition, the mix of Buddhism and shamanism in Oka creates specific types of knowledge, based on direct interactions of local residents with the landscape. Both Buddhist and shamanist beliefs are closely connected to the sacred view of Oka's landscape, where its different features have their own masters, or spirits (Ayusheev 2017). In Oka Buriat and Soiot shamanist beliefs, the landscape is seen as a sacred entity, various parts of which—including subsurface resources—are animated and governed by spirit masters (Gulgenova 2010). Oka Buriat and Soiot beliefs focus on a pantheon of celestial beings [*tengri*] where each god or spirit is associated with a specific natural force or part of the landscape (such as a river or a mountain). The sky is seen as a masculine force, which gives life, whereas the earth is associated with a feminine force, which shapes the elements. The heavens and the underground world are governed by strong spirits: *Erlen khan* [Mon.] is presented as the master of the underground world, including natural resources. Therefore, when extracting resources from the earth, one is at the same time negotiating with the spirit masters, who may allow or prohibit extraction. The sacred views on stones are represented through the cult of *oboo* [Mon.]—stone pyramids which are used for leaving offerings for spirit masters governing a particular territory (see Figure 10.2).

The local population of Oka is often characterized as people living in harmony with natural cycles (Shodorova and Iurieva 2010). The harsh climate of Oka, which did not allow for agriculture, resulted in the preservation of the traditional ways of subsistence of Buriats and Soiots, such as nomadic pastoralism and hunting (Pavlinskaia 2002). The herders who engage with yaks or reindeer follow the animals' routes and move together with them to the mountains for summer and then back to the valleys for winter. February-March is the time for preparing wood for the next winter, as frozen trees are easier to cut, and after cutting there is plenty of time for them to dry over the summer. In August, Orlik and other larger settlements become empty, as most locals leave for haymaking (if the grass is cut on time, it is believed to be more filling for the cattle). In autumn, the hunting season starts, and in winter, locals make trips to cities to sell the meat of wild animals and cattle (Interview B27; see also Oehler, chapter 12, this volume for an analysis of multispecies rhythms and cycles in Oka).

As the local residents of Oka largely depend on nature, it is important to follow its rhythms to sustain their daily lives. In an interview with the former Head of Okinskii District, Bair Baldanov,[5] this practice of adjusting daily

Figure 10.2 Oboo Near Khuzhir in Oka. *Source*: 2016, A. Varfolomeeva.

routines with the natural cycles was called "synchronization" [Rus. *sinkhronizatsiia*].[6] This concept resonates well with the way of local interactions with nature, which are based on non-hierarchical connections. The concept of synchronization implies a focus on coexistence which improves the overall performance of all the actors involved. In Oka, synchronization with the rhythms of nature is crucial for the local livelihood, as natural conditions may be rather harsh. In their daily routines, residents of Oka encounter floods and rock falls, extremely cold winters or rainy summers, as well as sudden weather changes. Therefore, it becomes especially important to be attentive to changes in nature.

A case from my fieldwork in Oka illustrates how the practices of synchronization with the rhythms of nature result in the entanglements of humans and nonhuman entities. During the drive from Orlik to Khuzhir, a small settlement in the northern part of the district, our car needed to pass a narrow road next to the mountain. Our guide Aleksandr announced cheerfully, "This is quite a dangerous part of the trip; the stones may fall onto the road at any moment." The car stopped, and Aleksandr asked us to join the ritual of making offerings to the local spirit masters with rice and drops of vodka. As this was the only road connecting Khuzhir with other settlements in Oka, I asked Aleksandr how the locals managed to pass here every day without being afraid of the possible rock fall. "Well, you just need to be prepared to take a risk,"—he answered. In situations like this, making offerings is viewed as a way of negotiating with the spirits of the territory (and, therefore, with nature in general). As it is not possible to predict or prevent rock falls, synchronization becomes a process of searching for cooperation with natural forces.

In Oka, the actions of humans and nonhuman entities are not organized in hierarchical order, but rather complement each other. By making offerings to the spirits, people acknowledge their power and ability to shape human lives. However, they are at the same time exercising their agency and looking for room to negotiate the future. Therefore, humans are not passive and obedient followers of nonhuman entities, as they exercise their agency by direct negotiation (e.g., making offerings, performing rituals, following established traditions). Resource extraction in Oka becomes an important part of this flexible cooperation between humans and the sacred landscape. The next part of the chapter will focus on the role of mineral resources in the process of negotiation between humans and spirit masters.

RESOURCESCAPE AND SPIRITS

The symbolism of stones and metals plays an important role in Buriat mythology, and it is closely connected to the entanglements formed between the landscape and its dwellers. In Buriat tradition, blacksmiths [Bur. *darkhan*] were compared to shamans in their power (Badmatsyrenova and Budain 2017, Lygdenova 2013). Blacksmiths were highly respected in Buriat communities, and it was believed that the craft of smithery was of celestial origin, sent to earth from the sky by the gods themselves (Sagalaev and Gombozhapov 2013). Blacksmiths were said to receive their power from working with fire, which has a purifying power according to Buriat tradition (Tsydenov 2012). Blacksmiths were viewed not just as workers, but as artists, much like painters or sculptors, as they were creatively changing the material world (Tsydenov 2012). Stones and metals also played an important role in the culture of Evenki of Buriatia. As one of my interviewees—a teacher of Evenki language in Ulan-Ude—mentioned, Evenkis used to work both with gold and jade and have legends relating to stone and metals (Interview B32).

An interesting metaphoric description of human-resource relations can be found in the Buriat epic poem, Geser, which was fully recorded in 1917 and first published in 1930 (Chakars 2014). As the events of the epos are believed to take place in the territory of the contemporary Okinskii district, Oka is often referred to as "the Land of Geser." The main character of the poem, Geser-Khan, is a celestial being who descended to earth in order to bring people well-being. In the fifth song of the poem, Geser is fighting with four evils which he aims to turn into four riches in order to help people.[7] The first evil takes the form of a turbulent river, which runs from the heights of Saridak Mountain on the border of Buriatia and Mongolia. The second evil appears in front of Geser-Khan as an aggressive dog, made of silver. Geser orders the dog to fall apart and to fill the pockets of people as silver, helping the poor. The third evil whom Geser meets is a dog made of pure gold, which tries to attack him with its shiny golden fangs. Nevertheless, Geser is not afraid. He tells the dog to "fall apart as pure gold" becoming "an insane passion for people." The final, fourth evil whom Geser needs to fight, is a mysterious one. It lies underground, unseen by people, and it hides its face hoping to go unnoticed by Geser. However, Geser manages to find the fourth evil and gives it an order using his magical powers: "Stop being a hidden evil, become eternal richness for people."

This part of the poem represents the interconnectedness of the landscape and its resources in Buriat beliefs. Out of the four evils, one (the river) represents natural forces, and the three others represent different sorts of natural resources (silver, gold, and other subsoil resources). It seems unclear from the poem which stone or metal the fourth evil represents, but most probably this is a common depiction of natural resources which are underground, hidden from people's eyes. All four entities (river, gold, silver, subsurface resources) are presented as "evils," which harm people, although they all have the potential of bringing them well-being. With his power and wisdom, Geser turns them from destructive forces into helpful substances: instead of danger, the river now offers healing, whereas silver, gold, and subsurface resources bring people economic benefits. However, their destructive nature is not completely eradicated by Geser, as he calls the gold "an insane passion." Therefore, natural resources are presented by the poem as a help and a danger at the same time.

In the Geser poem, natural resources are presented as actors with their own agency, but simultaneously as integral parts of the landscape, governed by spirit masters. This chapter draws on the notion of the *resourcescape* as an overarching concept that defines the interplay of different forms of resources in interactions between landscapes and humans. As Ey and Sherval (2015) note, resource extraction could therefore be understood as the process which reforms not only the physical environment, but also changes the sociocultural processes surrounding a mining site. The concept of resourcescape unites the notion of resource (including various forms of extraction such as mining, hunting, fishing, or berry picking) and the adaptability and versatility of the suffix "scape" (Soja 2006), implying the fluid nature of resource-rich landscapes. By using the concept of resourcescape, this chapter stresses that the entanglements of mining and landscape do not happen in isolation, but that they are part of the larger notion of human-resource relations. The concept of resourcescape demonstrates that in mining and other forms of extraction, such as hunting or fishing, humans and nonhuman entities are in mutual interaction, forming a common pluriverse that unites various perspectives and ontologies (Escobar 2011).

PAST AND PRESENT MINING IN OKA

The local residents of Oka have historically engaged with stoneworking, which still serves as a part of local legends and narratives. In the mid-nineteenth century, Soiot hunters accidentally found graphite of a high quality near Mount Botogol (Dorzhieva 2014). During this period, graphite was widely used for pencils and crucibles, while in Europe it was becoming a scarce commodity (Hobta and Snopkov 2007). Soon after graphite was discovered at Botogol, Jean-Pierre Alibert, an entrepreneur of French origin living in Russia, started a factory extracting the graphite. Alibert brought a team of French workers to help him, but he also employed local residents at the graphite mine. The extracted graphite was transferred to Germany, to the Faber Company in Nuremberg. Transportation of the extracted graphite was expensive, and thus the mine was closed in 1857, as a cheaper technology of producing graphite had been discovered in France. Soon after, Alibert left Oka, and the Botogol mine was abandoned. In the Soviet period, there were attempts to restart graphite extraction at Botogol (as graphite was used by the military). At the beginning of the 1990s the extraction stopped again (Hobta and Snopkov 2007) and during my fieldwork the locals believed it would not be continued in the near future (Interview B15).

Okinskii district is rich in mineral deposits including gold, jade, graphite, bauxite, phosphate, and asbestos. Locals often jokingly say that they have "all of Mendeleev's [Periodic] table under their feet." In the Soviet period, Oka became a site of active geological explorations. The first geological expeditions of state gold mining enterprise *Soyuzzoloto* in Oka took place in 1934 and 1938 (Akha, February 16, 2000). Geological brigades were searching for gold, jade, wolfram, molybdenum, and other natural resources. A number of local residents of Oka worked for geological brigades as guides. In 1955, one of the largest gold deposits of the district, Zun-Kholba, was discovered. Today, gold mining is primarily carried out by *Buriatzoloto*. The company makes up around 70 percent of Oka's budget by way of its revenue (Interview B6).

Jade was discovered in the Eastern Saians in the middle of the nineteenth century, but it did not receive much attention from the state. The stone was merely used in the decoration of buildings or for interior details. The search for jade began in the Soviet period, and several deposits were discovered in Buriatia before the Second World War (Inform Polis, May 30, 2018). In the 1980s, jade was discovered in Oka. Since the 1990s, several jade deposits in Oka have been controlled by the company *Baikalkvartssamotsvety* based in Irkutsk.[8]

In Russia, jade has no special value and is classified as decorative stone (and not semi-precious). In China, however, it is considered a sacred stone, and in certain historical periods it was valued higher than gold or silver. Jade has a strong symbolic meaning in China, being connected to wealth, power, and social class (Hsu et al. 2015). Since the early 2000s, the demand for jade increased in China due to the Beijing Olympic Games held in 2008. Beijing gold, silver, and bronze Olympic medals were mounted with different types of jade (white jade for gold medals and green jade for silver and bronze medals). Prior to the Olympic Games, the Chinese government launched an educational and promotional campaign for jade, highlighting its significance in Chinese history and culture. Due to the campaign, the price of white and green jade increased in China. Buriatia, where the gemstone could be bought more cheaply, became one of the main markets for jade. Chinese traders were searching for all sorts of jade, even for less valuable grades, in order to sell them to tourists during the Olympic Games (Safonova and Sántha 2010a). During this period, informal jade extraction became widespread in Oka (Interview B10), as well as in other regions of Buriatia, such as Baunt District, which is rich in white jade. Nowadays, informal jade extraction is a widespread occupation in Oka among young and middle-aged men (Interview B4). In most cases they travel to Kitoi or Khara-Nur, two main jade deposits in the district. The extracted jade is then brought to Irkutsk or Ulan-Ude and sold to Chinese traders whose contacts are shared by word of mouth (Interview B8). The residents of other neighboring regions (e.g., Tunka district) also often travel to Oka to extract jade (Interview B5).

Jade trips often involve infrastructural challenges. As forest paths are difficult to navigate, locals use four-wheelers or all-terrain vehicles, such as UAZ or KamAZ (both are types of Russian-made off-road vehicles). This is a risky occupation, as in many cases local residents, while trying to collect stones left from previous extractions of *Baikalkvartssamotsvety* enter into direct conflict with the company's guards. During the New Year night of 2012, five local residents tried to steal pieces of jade from a warehouse and were wounded by the guards (Novaia Buriatia, January 25, 2012). In 2014, as a result of a conflict among jade miners from different regions at one of the Oka's deposits, two people were killed and four wounded (Inform Polis, June 1, 2017). In August 2015, during the attempt to steal jade from a warehouse, one of the local residents of Oka was killed (ibid.).

During my fieldwork, many locals were worried that the official status of jade would change (from decorative to semi-precious), and that this change would result in more serious punishment of informal miners. They were not sure if they would have opportunities for informal jade mining in the near future, especially after the conflicts of 2012 and 2015. One of the interviewees, Tatiana, stated, "Putin is still not sufficiently interested in our jade, but if he gets interested, everything will stop here" (Interview B16). Another local resident, Damba, believed that private businesses would soon strengthen their control over jade deposits in the district: "Anyway big companies, those who have money, buy everything" (Interview B15). The escalating conflicts between private companies, informal miners, and the state trying to control the extraction, raised concerns of resource entitlement and ownership. My interlocutors in Oka often questioned the right of mining companies, most of which were from outside the region, to extract resources in their territory. On the contrary, they saw local residents to have the right to extract resources, as resources belonged to those who reside on a particular land (Interview B15). Residents claimed that many locals had to become illegal miners, as this was the only way to survive in a setting where there existed no opportunities to become legally involved with jade extraction (Interview B5). This theme of ownership "by law," which contradicted symbolic ownership relating to territorial rights, often came up in the interviews. My interviewee Nomina, for example, sadly said that, just as Soiot hunters had to share their local graphite with entrepreneurs in the nineteenth century, today's residents of Oka needed to share their resources with investors and private companies, even if they had found it first (Interview B27). In this sense, informal jade mining in Oka became a way to question established ownership schemes and to demonstrate the right of local residents to extract natural resources.

JADE IN OKA'S RESOURCESCAPE

This section of the chapter demonstrates that jade extraction in Oka cannot be viewed separately from other resources in the district, such as hunting or the land itself. Although informal jade mining in Okinskii district represents a relatively modern phenomenon, it has become embedded in the established relations of synchronization existing between humans and nonhuman entities. The local way of searching for coexistence with the natural forces of the landscape also has its implications on informal jade extraction, which becomes a part of negotiations between humans and spirits. The process of jade extraction is seen as taking something from the land and from its spirits, and thus it has to be performed carefully and with respect toward the masters of the territory.

A number of ethnographic studies draw parallels between stone extraction and other forms of interaction with nature. Margaret Moore defines the term "natural resources" in an open way and uses it to refer to "anything, derived from the environment and not made by humans that is instrumental to satisfying human wants and needs" (Moore 2017, 163). Land, water, and sunshine are resources, as are plants, animals, and mineral ores. Safonova and Sántha (2016) find similarities between jade extraction and the gathering of ice (which is the main source of drinking water in winter) among the Evenki in Buriatia. Jade and ice are similarly collected in pieces and stored on wooden platforms, after which ice is used for water while jade is transferred to Chinese traders. Both melting ice and sold jade symbolically disappear from the local community. Another study by Safonova and Sántha (2010b) discusses the similarities between Evenki hunting style and the way of doing business when extracting jade in their territory. Jessica Graybill (2009) suggests that hunting in the Russian East can be viewed historically as an extractive industry, a precursor to economic practices organized around resources. Büscher and Davidov (2013) demonstrate that ecotourism and extraction in many cases share a similar logic and are often experienced by locals in a similar way. Extraction and ecotourism, as Büscher and Davidov show, are more similar than is often imagined: as they are related not only through the experiences of the people they affect, but also through the broader structures of power and ownership over resources.

It is possible to draw parallels between jade and other types of resource extraction. I refer to the jade trips as "informal mining," but jade trips do not always involve the actual extraction of the stone (Interview B4). In many cases, locals do not extract jade, but rather take bits and pieces which may remain at extraction sites. In this sense, their activity could be referred to as "jade picking," similarly to berry and mushroom picking or other types of foraging. Even though interviewees referred to their activities as "stealing" (Interview B8), they did not see this as a problematic issue, as jade was not believed to be "owned" by the extractive businesses even though they may have had an official license. Jade was perceived as a part of nature, similarly to animals, fish, or plants. Consequently, it seemed logical to take some to satisfy one's needs. One of my interviewees, when recalling his experience in illegal mining, said that he did not see any problems in taking something that belonged to him anyway (Interview B4). These notions of jade as a "shared resource" in Oka correspond to similar perceptions among Tozhu hunters who view resource extraction, whether hunting or sharing gasoline with the industrial workers, as reciprocal exchange (Jefanovas, this volume).

During my fieldwork in Buriatia, when I listened to the stories about jade extraction trips, such narratives often resembled the way the residents of Oka talked about hunting and the rituals related to it. Just as the right balance between hunters, landscape, and spirit masters of the territory was important for hunting (Oehler 2018), jade extraction trips were viewed as successful when a similar balance had been achieved. There are several possible parallels between jade extraction and hunting in Oka. First of all, both jade and hunting trips involve a strong element of luck. Just like hunters, informal jade miners never knew how they would return from the forest. If they were successful and managed to sell their pieces of jade, they might make enough money to buy a new car or renovate their house (Interview B5, B8). My interviewee Boris, who used to work in Oka, was certain that the recent appearance of expensive cars was most probably related to informal jade extraction (Interview B25). Another interviewee, Andrei from Orlik, told me directly that his jeep was a result of a successful jade trip (Interview B8). During the fieldwork I was also told that it was possible to differentiate between fences surrounding the houses in Orlik: the men from the households with expensive-looking fences were most probably engaged in illegal jade mining.

In some cases, nature prevented informal miners from reaching the stone. Jade extraction trips also largely depended on natural conditions. Andrei (Interview B8) shared a story of his truck getting stuck in the snow during one of the trips. Eventually Andrei's crew had to turn back. He and his friends saw the heavy snowfall as a sign from nature itself [Rus. *sama priroda*] forbidding them to reach their destination (Interview B8).

At the same time, all the investments made by a crew may be in vain if the piece of extracted jade turns out not to be very valuable. Balta from Orlik (interview B5) had experienced an unsuccessful jade extraction trip when he and his friends were stopped by police on their return, pretending they were carrying fish instead. The policemen did not believe Balta and his friends, searched their car, and found several pieces of jade (Interview B5). As Balta knew one of the policemen personally, they agreed that the police would let them go if they left all the jade they took. They threw the jade pieces away, but remembered the place and returned the next morning to retrieve them. In the end, however, the pieces they had mined turned out not to be valuable, obtaining only around 1,000 rubles[9] each. The story of Balta demonstrates that during jade hunting trips the local residents need to enter into a complex set of negotiations not only with nonhuman entities, but with the representatives of the state as well.

As Broz and Willerslev (2012) note, the concept of luck in Siberian hunting societies differs from its established Western counterpart (as chance or coincidence). Indigenous views on hunting luck are closely linked to local spirituality. A lucky hunt depends on the goodwill of the animal's spirit owner and therefore needs to be negotiated (Willerslev 2004). Similarly, the eventual success or failure of jade extraction trips are viewed by the residents of Oka as the result of successful or unsuccessful negotiations with the spirit masters of the territory. In order to secure the goodwill of the spirits, it is important to respect the stone and not to be too greedy for it. Ziker (2003), in his study of hunting traditions in Taimyr, stresses the dangers of overhunting, which is seen as breaking the moral code of negotiating with the spirits and therefore may be punished. In Oka's informal jade extraction sphere, it is considered appropriate to get enough jade to secure one's well-being, but people should not try their luck and search for more, as it will result in misfortune. As Anna Tsing notes in her study of a "resource frontier" in Indonesia, natural resources "become fugitive in this landscape of movement and flight . . . even safe in one's possession, treasures disappear" (Tsing 2003, 5102)

The fugitive nature of natural resources which may be given to people or taken from them depending on the will of the resource's spirit master is visible in Oka. Several of my interviewees stated that jade had a strong energy and therefore would take revenge on those who did not respect it. Nikanor from Orlik told me that jade punishes those who are greedy. He illustrated this thought with a story about his neighbor who could not stop after a successful jade trip and continued going out for more. Eventually, this greed had resulted in a series of misfortunes for the miner and his whole family (Interview B7). Another interviewee, Viktor, stated that jade could bring too much money too fast, and this disproportionate profit was risky for the local culture [Rus. *mental'nost'*] of the Oka community (Interview B25). He believed, the local identity of Oka Buriats and Soiots may be threatened by the enormous fortunes promised by informal jade extraction. Instead of keeping strong connections with the territory, locals would start thinking primarily about profit making.

Both hunting and jade extraction are associated with multiple layers of danger. Losing the goodwill of the spirit masters represented one of the dangers in hunting, as this could result in illnesses or other misfortunes for a hunter's family (Broz and Willerslev 2012). Another source of danger in hunting was wild animals: one of my interviewees recalled a story about his uncle encountering a bear while hunting in Oka (Interview B11). One of the common dangers associated with jade extraction was the possibility of being stopped by police and getting arrested (Interview B5). Another even more serious danger was related to cases in which locals tried to take the pieces of jade stored in the warehouses of private mining companies. Tatiana from Sorok complained, during our interview, that locals had to walk under flying bullets to obtain the stone. She was referring to the episode from August 2015, when a young resident of Orlik was accidentally shot by the guards while trying to steal pieces of jade from the company's warehouse (Interview B16). After this tragedy, the number of active informal miners dropped, but many locals were still willing to risk much. "What else can we do? There are no jobs here, only cattle and jade," said Nomina, a resident of Sorok, during our interview (Interview B27).[10]

Due to the high risks associated with extraction, jade trips were organized in secret, and plans were kept even from relatives. As Andrei expressed it, "I did not tell my family we are going for jade, as they would have been worried" (Interview B8). It seemed, though, that partially due to this danger, jade extraction was appealing for young people: jade represented "fast money" and could make its owner rich overnight. At the same time, it was risky and required courage and adaptability (Interview B11). In Orlik, I met a group of young men who told me happily, "You can ask us anything about jade, jade is our life!" showing me photos and videos of their jade trips from a mobile phone. In one of the videos, the young men were crossing a rapid stream on four-wheelers. In another, they demonstrated a large piece of jade they had managed to find. These stories resembled hunting narratives in which the hunters demonstrate their bravery and the amount of prey they got.

The local residents of Oka did not see much value in jade per se, though it was sometimes used in the sauna [Rus. *bania*], as jade steam was believed to be good for health (Interview B10). Some residents of Oka had pieces of jade at home, but they were seen more like a trophy or a souvenir and were not kept for their visual appeal. These were pieces that had been found or extracted during a jade trip, but which had not been of sufficient value to be sold. Pieces of jade kept at home were normally hidden and not shown openly. Therefore, demonstration of a piece of jade was a sign of trust, and these "hunting trophies" were seen as a way to prove miners' bravery, courage, and luck. As the jade trips were often linked to risk taking, keeping the pieces of jade at home signified the owner of the jade as an individual who managed to overcome risks and challenges and did not return empty-handed.

As the pieces of jade were kept in secret, and informal mining itself was an activity rarely spoken about openly, the perception of jade in Oka was truly complex. Locals did not praise jade for its looks or physical qualities. Instead, they valued it as a source of income. "Why would these Chinese even be interested in jade, it seems nothing special," one of my interviewees reflected (Interview B10). On the other hand, as a part of the sacred landscape, jade was seen as a powerful actor, which could "punish those who are greedy" or take their luck away. It was therefore important to negotiate with the stone and its spirits in order to ensure "jade hunting luck." Jade also strongly influenced people's lives. As Nomina stated, many locals needed to engage in jade mining despite its dangers if they were to have a job and stay in Oka. Becoming embedded in previously existing connections between humans and the landscape, jade was reinforcing these same relations, while creating new links and entanglements. Because of their engagement in informal jade extraction, local residents were able to stay in the district, keeping strong connections with the land and its spirit masters.

CONCLUSION

The resourcescape of Okinskii district reflects a complex entanglement of various forms of natural resource extraction. Informal jade mining, which began relatively recently in Oka, has become embedded in previously existing forms of subsistence hunting, following a similar logic. The concept of luck becomes one of the organizing principles of jade extraction trips. Just as in hunting, in informal jade mining it becomes crucial to respect the resource and its spirit masters, to be modest and not to take too much. Small pieces of jade may be kept at home, resembling "hunting trophies." They are viewed as reminders of previous trips, the risks associated with the activity, and, finally, of strength and masculinity. The line between various forms of extraction, such as jade mining and hunting, is therefore blurred, even though these activities are clearly divided into "legitimate" and "illegal" in the eyes of the state. Both informal jade mining and other forms of extraction are deeply rooted in complex perceptions of sacred landscape in Oka and in the set of relations between humans and territory masters.

Jade may be viewed as a force influencing people's lives in Oka: my interviewees felt they had to preserve their connections with the stone were they to stay in Oka, given the region's limited opportunities for employment. At the same time, local residents took an active role in negotiating with the stone and its spirit masters. As they synchronize their movements and activities with various natural rhythms and cycles, jade extraction trips become a part of this synchronization process. In this sense, the relations between humans, non-human sentient beings, and materials could be viewed as cohabitation (see Rasiulis, this volume). As Hamayon (2012) notes, hunting luck depends on the active behavior of the hunters and is contingent on complex interactions between various types of actors. This could be well applied to the informal jade extraction in Oka. The extraction of jade becomes a complex interplay between miners, spirits, landscape, but also the state as the controlling force, which tries to shape or limit the extraction.

Jade trips in Oka, therefore, could be seen as a process involving multiple negotiations: with the spirit masters of the territory, as well as with the state. Balta's example of pretending that he and his friends were fishermen represents the process of negotiations between the state (as the police) and the citizens. Informal jade extraction activities in Oka were perceived as a realization of the symbolic right to extract local resources "from one's own land." Even though the power of the state was well recognized in Buriatia, people found ways to temporarily shift established power relations. This was accomplished through micro-actions, such as promoting one's special connection to the land and its resources, or through activities directly contradicting the state and its policies, such as informal jade extraction and transportation.

The notion of "resourcescape" used in this chapter refers to blurred borders between different types of resource extraction in Oka. At the same time, it also refers to the strong connections of all types of extraction to the land and its dwellers. The concept of resourcescape demonstrates that the notions of nature and industry may be perceived as mutually entangled concepts which do not stand in opposition to each other. In Oka, graphite, gold, and jade mining are deeply embedded in local histories and narratives. These narratives include legends about the technological advancements of Alibert's mine, interactions with Soviet-era geological brigades, and shared stories about the power of jade to punish greedy miners. Through local relations to mining, the notions of rootedness and symbolic ownership are made evident. Being rooted in Oka meant having rights to participate in resource extraction. Therefore, local connections to jade became an integral part of the sacred relations between humans, landscape, and spirit masters, and the deep knowledge of these relations gave local residents symbolic ownership over their resources.

List of cited interviews: B4: male, b. 1961, Orlik; B5: male, b. 1986, Orlik; B6: female, b. 1989, Orlik; B7: male, b. 1960, Orlik; B8: male, b. 1984, Orlik; B10: male, b. 1988, Orlik; B15: male, b. 1958, Sorok; B16: female, b. 1955, Sorok; B25: male, b. 1954, Ulan-Ude; B27: female, b. 1986, Sorok; B32: female, b. 1957, Ulan-Ude

NOTES

1. Here, I use district as a translation for the Russian word raion which is a type of administrative unit of several post-Soviet states, usually a subdivision two steps below the national level.

2. Oka (or, in Buriat language, Akha) is the informal name of Okinskii district.

3. The interview indicators consist of B (for Buriatia) and the order number of the interview.

4. Akha (previously Saianskaia Nov') is the newspaper of Okinskii district.

5. Bair Baldanov was the Head of Okinskii district at the time of my first fieldwork in Buriatia in summer 2016. In autumn of 2016, he resigned and moved to Ulan-Ude to take the post of the editor of Buriad Unen newspaper.

6. Interview with Bair Baldanov: *Selo Rodnoe*, January 13, 2016: http://selorodnoe.ru/projects/show/id3694542/ (accessed March 20, 2019).

7. All the quotes are translated from the Russian version of Geser (translated from the Buriat language by Vladimir Soloukhin and Semen Lipkin).

8. *Baikalkvartssamotsvety* was founded in 1966 in the Irkutsk region of Russia as a state enterprise. In 2001 the state enterprise was bankrupt and became privatized in 2003 (Tunka24, January 15, 2016).

9. At the end of 2014, when Balta's jade trip took place, 1,000 rubles equaled around 20 dollars.

10. As a female, Nomina was not directly involved in jade extraction, but knew many residents of Sorok who used to engage in jade mining in the past or at the time of our interview in 2016.

REFERENCES

Ayusheev, Norbu. 2017. "Soioty Buriatii v Dukhovnoi Tsivilizatsii Saianidov Vnutrennei Azii" [Soiots of Buriatia in the Sacred Civilization of Inner Asia Saianides]. *Vestnik Buriatskogo gosudarstvennogo universiteta. Gumanitarnye issledovaniia Vnutrennei Azii*, 1: 86–92.

Badmatsyrenova, Darima, and Ariuna Budain. 2017. "Elementy Shamanizma v Buriatskom Epose o Gesere" [The Elements of Shamanism in Buriat Epic Poem about Geser]. *Filologicheskie nauki. Voprosy teorii i praktiki*, 1–1 (67): 65–68.

Belyaeva-Sachuk, Veronika. 2012. "Soioty – Molodoi Sibirskii Etnos. Primer Sozdaniia Novoi Etnicheskoi Identichnosti v Sovremennoi Rossii" [Soiots – a Young Siberian Ethnic Group. An Example of Creation of a New Ethnic Identity in Contemporary Russia]. *HOMO COMMUNICANS: Człowiek w Przestrzeni Komunikacji Międzykulturowej II*. Uniwersytet Szczeciński: Wydział Humanistyczny.

Broz, Ludek, and Rane Willerslev. "When Good Luck is Bad Fortune: Between Too Little and Too Much Hunting Success in Siberia." *Social Analysis* 56.2 (2012): 73–89.

Bubandt, Nils. 2017. "Haunted Geologies: Spirits, Stones, and the Necropolitics of the Anthropocene." In *Arts of Living on a Damaged Planet: Ghosts and Monsters of the Anthropocene*, 121–141. Minneapolis: University of Minnesota Press.

Büscher, Bram, and Veronika Davidov. 2013. "Introduction: The Ecotourism-Extraction Nexus." In *The Ecotourism-Extraction Nexus: Political Economies And Rural Realities of (Un)Comfortable Bedfellows* edited by Bram Büscher and Veronica Davidov, 1–16. London: Routledge.

Chakars, Melissa. 2014. *The Socialist Way of Life in Siberia: Transformation in Buryatia*. Budapest, New York: Central European University Press.

Dorzhieva, Galina. 2014. "Pik Alibera (Frantsiia) i Doroga Alibera (Rossiia, Buriatiia) – Vremen Sviazuiushchaia Nit'" [The Peak of Alibert (France) and the Road of Alibert (Rissia, Buriatia)—the Thread Connecting Times]. *Vestnik Irkutskogo Gosudarstvennogo Universiteta* 7 (90): 190–194.

Escobar, Arturo. 2011. "Sustainability: Design for the Pluriverse." *Development* 54 (2): 137–140. doi:10.1057/dev.2011.28.

Ey, Melina, and Meg Sherval. 2015. "Exploring the Minescape: Engaging with the Complexity of the Extractive Sector." *Area* 48 (2): 176–182. doi:10.1111/area.12245.

Graybill, Jessica. 2009. "Places and Identities on Sakhalin Island: Situating the Emerging Movements for 'Sustainable Sakhalin.'" In *Environmental Justice and Sustainability in the Former Soviet Union*, edited by Julian Agyeman and Yelena Ogneva-Himmelberger, 71–95. Cambridge, MA: MIT Press.

Gulgenova, Sendyma. 2010. "Sovremennyi Etnopsihologicheskii Oblik Soiotov i Saianskih Buriat" [The Contemporary Ethnopsychological Image of Soiots and Saian Buriats]. *Vestnik Buriatskogo Gosudarstvennogo Universiteta. Biologiia. Geografiia* no. 4: 35–38.

Hamayon, Roberte N. 2012. "The Three Duties of Good Fortune: 'Luck' as a Relational Process among Hunting Peoples of the Siberian Forest in Pre-Soviet Times." *Social Analysis* 56 (1). doi:10.3167/sa.2012.560107.

Hobta, Aleksandr, and Sergei Snopkov. 2007. "V Gostiakh u Alibera" [Visiting Alibert]. *Zemlia Irkutskaia* 1 (32): 48–59.

Hsu, Tao, Kirk Makepeace, Andrew Lucas, and Nikki Makepeace. 2015. "The Nephrite Jade Road: Evolution of the Green Nephrite Market | Research & News." *Gia. edu*. Accessed May 2, 2019. https://www.gia.edu/gia-news-research/nephrite-jade-road-evolution-green-nephrite-market.

Ingold, Tim. 2007. "Materials Against Materiality." *Archaeological Dialogues* 14 (1): 1–16. doi:10.1017/s1380203807002127.

Lygdenova, Viktoria. 2013. "K Voprosu o Kul'te Kuznetsov u Barguzinskikh i Kurumkanskikh Buriat" [On the Question of Blacksmiths Cult among the Buriats of Barguzin and Kurumkan]. *Vestnik Tomskogo Godusarstvennogo Universiteta. Istoriia* 2 (22): 62–66.

Mongush, Marina. 2012. "Tofalary i Soioty: Istoriko-Etnograficheskii Ocherk" [Tofalars and Soiots: a Historical and Ethnographic Essay]. *Novye issledovaniia Tuvy* 2 (14): 62–78.

Moore, Margaret. 2017. *A Political Theory of Territory*. Oxford: Oxford University Press.

Oehler, Alexander. 2018. "Hunters in their Own Right: Perspectival Sharing in Soiot Hunters and their Dogs." In *Dogs in the North: Stories of Cooperation and*

Co-Domestication, edited by Robert J. Losey, Robert P. Wishart, Jan Peter Laurens Loovers, 28–44. London: Routledge.

Pavlinskaia, Larisa. 2002. *Kochevniki Golubykh Gor. Sud'ba Traditsionnoi Kul'tury Narodov Vostochnykh Saian v Kontekste Vzaimodeistviia s Sovremennost'iu.* [Nomads of the Blue Mountains. The Fate of Traditional Culture of the Peoples in the Eastern Saians in the Context of their Interactions with Modernity] St. Petersburg: Evropeiskii Dom.

Safonova, Tatiana, and István Sántha. 2010a. "Being Local Outsiders: A Study of Chinese Ethos in East Siberia." *Inner Asia* 12 (2): 347–364.

———. 2010b. "Different Risks, Different Biographies: The Roles of Reversibility for Buryats and Circularity for Evenki People." *Forum Qualitative Sozialforschung / Forum: Qualitative Social Research* 11(1).

———. 2016. "Gathering as Ontological Practice among Evenki of Eastern Siberia." *Slovak Ethnology* 2 (64): 192–227.

Sagalaev, Konstantin, and Arsalan Gombozhapov. 2013. "Kuznets i Shaman" [Blacksmith and Shaman]. *Nauka iz Pervykh Ruk* 5–6 (53–54): 84–95.

Shodorova, Nina, and Nina Iur'eva. 2010. "Organizatsionno-Ekonomicheskie Mekhanizmy Razvitiia Ekologicheskogo Turizma na Primere Okinskogo Raiona Respubliki Buriatiia" [Organizational and Economic Mechanisms of Ecological Tourism Development on the Example of Okinskii District of the Republic of Buriatia]. *Vestnik Buriatskogo Gosudarstvennogo Universiteta* no. 2: 70–75.

Soja, Edward W. 2006. "Foreword: Cityscapes as Cityspaces." In *Urban Space and Cityscapes: Perspectives from Modern and Contemporary Culture*, edited by Christoph Lindner, xv–wviii. London: Routledge.

Tsing, Anna Lowenhaupt. 2003. "Natural Resources and Capitalist Frontiers." *Economic and Political Weekly* 38(48): 5100–5106. doi: 10.2307/4414348.

Tsydenov, Enhe. 2012. "K Nekotorym Obriadam v Sovremennom Buriatskom Shamanizme." [On Certain Rites in Modern Buriat Shamanism]. *Vestnik Buriatskogo Gosudarstvennogo Universiteta. Pedagogika. Filologiia. Filosofiia* 8: 242–247.

Vakhtin, Nikolai, Evgenii Golovko, and Peter Schweitzer. 2004. *Russkie Starozhily Sibiri: Social'nye i Simvolicheskie Aspekty Samosoznaniia.* [Russian Old-Settlers in Siberia: Social and Symbolic Aspects of Identity]. Moscow: Novoe izdatel'stvo.

Varfolomeeva, Anna. 2019. "Articulations of Indigeneity in Two Mining Regions of Russia: A Comparative Case Study of Karelia and Buriatia." PhD diss., Central European University.

Viveiros de Castro, Eduardo. 2004. "Exchanging Perspectives: The Transformation of Objects into Subjects in Amerindian Ontologies." *Common Knowledge* 10 (3): 463–484. doi:10.1215/0961754x-10-3-463.

Willerslev, Rane. 2004. "Not Animal, Not Not-Animal: Hunting, Imitation and Empathetic Knowledge among the Siberian Yukaghirs." *Journal of the Royal Anthropological Institute* 10.3 (2004): 629–652.

Ziker, John P. 2003. "Assigned Territories, Family/Clan/Communal Holdings, and Common-Pool Resources in the Taimyr Autonomous Region, Northern Russia." *Human Ecology* 31 (3): 331–368. doi:10.1023/a:1025024804641.

Chapter 11

Falling In and Out of Rhythm

Alex Oehler

Perhaps one of the most iconic aspects of pastoral Inner Asia is what in Mongolian is known as *tavan khoshuu mal*, or "the five kinds of animal" (Fijn 2011, 18). The phrase refers to hoofed herd animals, or *mal*, which in Mongolia commonly are horses, cattle, sheep, goats, and camels (ibid. 23). As one ventures further north into the mountainous taiga of southern Siberia, the composition, importance, and number of "the five kinds of animal" begins to vary. Most Oka-Soiot households of western Buriatia, which I will discuss in this chapter, comprised yak, Mongolian dairy cattle, horses, sheep, and goats, although goats were less common. Along with chickens, goats had only recently been adopted by several households as a means of "fast meat" [Rus. *bystroe miaso*]. According to their owners, goats grew swiftly and they were not picky eaters. Their meat quality was considered quite good.

More importantly, "fast meat" could fill a gap in the seasonal meat availability of a household. Given the high elevation and short warm period, people of Oka were reliant on meat and milk for their protein. Much like opportunistically hunted deer, which occasionally passed close enough to people's homes, goats offered a source of sustenance during periods of protein scarcity. Although people had year-round access to herds, these were used sparingly for their meat. Yaks were slaughtered only in autumn, and their meat was a precious source of cash. Horse herds produced precious transportation animals. And with electricity powered refrigeration being limited to winter residences, long-term preservation of large animal bodies during warmer seasons was impractical. Smaller animals, on the other hand, could provide meat for a larger family for several days without requiring preservation, other than cooking.

By May most Soiot households in Uro Valley near the village of Sorok had used up their winter meat supplies. This was also the time when residents were preparing to depart their winter pastures to head into the mountains to their summer pastures. Meanwhile, their dairy cows were having their calves, which left little milk for human consumption. With little milk and practically no meat, the sight of someone inviting their neighbors to partake in the slaughter feast of a goat was more than welcome. The timing of the goat—an otherwise non-essential animal to the Soiot household—could thus become quite strategic. The speed of its growth, which came at little extra cost to households, and its position as being always 'ready-at-hand,' were features that fit well with the rhythms of other more dominant and work-intensive species.

I provide the goat as an example to introduce the idea of species-specific speed and rhythm in Oka households. It could be argued that it was the uncomplicated and economically non-essential role of the goat that secured it a place in the Soiot household. More importantly, it's feeding and movement requirements did not interfere with the seasonal rhythms of other species constituting the household. Households sometimes had goats, other times they did not. To buy two goats was not a big deal, and it did not require much patience to build a little flock that could be kept near the encampment, while people focused on tasks relating to other, more demanding species. Goats fit a role and purpose that had existed long before their adoption into the household. In a way, they were interchangeable with fish and deer by helping bridge meat rich winters and milk secure summers.

Although anthropologists and scholars from cognate disciplines increasingly invoke post-humanist language, aiming for "an anthropology beyond the human" (cf. Rees 2018; Kohn 2013), the study of other-than-human perception remains rather a new territory within our discipline. In this chapter, I seek to invoke a discussion of how multispecies households manage differences in the speed perception and landscape preferences of their nonhuman members. How do the rhythms of animals, people, climate, and landscapes work together to produce ecologies worthy of inhabiting together? Given environmental and political changes, how do the relationships between species change within perpetually adapting households? In an attempt to answer these questions through the lens of ethnography from Oka, I will also draw on the help of German biosemiotitian Jakob von Uexküll (1864–1944) and French philosopher and sociologist Henri Lefebvre (1901–1991).

DOMESTICATION STUDIES IN INNER ASIA

It is difficult for anthropologists working in Inner Asia to bypass discussions of pastoralism, and therefore of animal-human relations. Human proximity to animals arguably permeates almost all spheres of society beyond the city in Inner Asia. It is not surprising then, that the study of the origins of animal domestication has been of interest to many scholars of Inner Asia. Much of the work of these scholars, as it pertains to the incorporation of animals into human households, contests conventional notions of animal domestication. Underpinning the concept of the Neolithic Revolution (cf. Childe 1951) was the idea that humans had gained unprecedented control over animals and the wider environment. Following this line of thought, the new found freedom from the unpredictability of other beings enabled our manipulation of the rhythms and life speed of other life forms.

Recent scholarly discontent with this streamlined depiction of domestication goes far beyond Inner Asia, of course. As Lien, Swanson and Ween (2018, 14–18) point out, much of it grows from recent ethnographic and archaeological documentation of animal-human relations from across the globe, involving past and present scenarios: increasingly, archaeologists are suggesting animal domestication may have been a process that required about four to five millennia, involving many regressive stages in which animals repeatedly went feral before becoming re-domesticated. Domestication would also have taken place in more than one place at the same time, each time with possibly different results. Most importantly, perhaps, domestication would not have been the result of human intention alone, but also of diverse co-evolutionary and mutualistic processes. In fact, many of the physiological changes that resulted in animals from their new living conditions, according to this scholarship, would have been unforeseen by humans.

Ethnographic work from Inner and North Asia supports many of these arguments. Anthropologists Stépanoff et al. (2017) and others (Anderson 2014; Cassidy & Mullin 2007) have described animal domestication as an ongoing process, a relationship between humans and other animals that continuously calls for reassessment and renegotiation by its participants. Stépanoff et al. refer to one facet of this process as "intermittent coexistence," which for them is a form of perpetual relational becoming in which not only human but also other-than-human autonomy is taken seriously. Conventionally, domestication and increased docility or tameness were seen as going hand in hand. Conversely, intermittent coexistence allows for periods of separation between people and animals, thereby encouraging a degree of independence from human care.

Much of this independence is directly rooted in allowing animals to adhere to their own reading or perception of the landscapes in which they subsist. In other words, species belonging to households are encouraged to follow their own rhythm and perception of the environment. This is not a far fetch: herders and hunters themselves are highly aware of the life rhythms that surround them. As we can see from recent ethnographic work with Evenkis of southeastern Siberia (Brandišauskas 2017), hunters and herders are keenly aware of the rhythms inherent in specific places, which are made visible or audible by the changing actions and behaviour of its animal inhabitants. People and animals mold their own rhythms—including decisions regarding timing and direction for relocation or migration—in relation to their knowledge of other rhythms.

It is well known to hunters and herders alike that animals adjust their actions, depending on their recognition of what American psychologist James J. Gibson calls affordances (1986 [1979]). While hunters are familiar with predictable sequences in the seasonal transformations of landscapes and animals—an example being the timing of reindeer as they descend to lower elevation in autumn to search for mushrooms—even the most skillful of hunters and herders rely on careful reading of animals to confirm exact timing. Knowledge of rhythms helps anticipate seasonal transitions, but attentive reading is still required to know when a transition has begun. Fine-tuned domestication relationships are subject to a similar kind of careful reading, further problematizing the conventional divide between hunting and herding activities.

Much like a hunter, who moves between the diverse life rhythms of prey animals, a herder attempts to reconcile the rhythms of multiple species. The herder—by way of the household—attempts to reconcile the rhythms of the species found within the household. Yet, this work may well include the reconciliation of rhythms belonging to species found both inside and outside of the boundaries of what is conventionally understood as a household. Because different species recognize non-identical affordances in shared landscapes, it can be a challenge to form and maintain a sustainable multispecies household. A key aspect of making successful homes together—whichever definition of the household we may hold on to—is the ability to accommodate the rhythms of those whose paths intersect regularly.

RHYTHM AND UMWELT

Although Henri Lefebvre wrote from an urban perspective, some of his ideas regarding time and space seem to lend themselves to the remote mountains of Oka. For him, change is much more a matter of repeating cycles than of a linear progression of events. A cyclical approach to time stands closer to people whose lives are carefully attuned to the seasons, and whose perception of time draws on the recurring, yet never perfectly predictable, actions of animals. In his close observations of social life, Lefebvre notes inorganic patterns produced by machines. Yet, his emphasis lies on the nature of rhythm as found in social and biological contexts. He attempts to shine a light on how social and biological rhythms influence each other in the formation of habits

(Lefebvre 2004[1992], 75). Habituation is one area in which rhythm and domestication come together. In fact, the formation of multispecies habits is variously discussed as integral to the domestication of animals (cf. Brubaker and Coss 2015; Candea 2013).

Furthermore, Lefebvre distinguishes between rhythm and sequence. He also makes a difference between rhythm and movement, or speed. Rhythm can emerge only where there is some form of movement that causes repetition. He calls this repetition measure, a term familiar to us from music. But what defines his concept of organic rhythm is the absence of "identical absolute repetition" (Lefebvre 2004[1992], 9). Organic measure, for him, remains in part unpredictable. It is this unpredictability that enables the coming together of repetition with difference or change. In Lefebvre's view, difference or change is perpetually introduced into existing rhythms. He describes this process as a coming together of qualitative and quantitative qualities, an example being the fusion of measurable time with spontaneous harmony. But of importance to our discussion is his observation that organic rhythm can be an agent of change.

To detect or to read the emergence of change in the rhythms of others, the rhythm analyst must allow her own rhythm to intersect with that of the subject. This makes the study of rhythms a highly somatic undertaking. The rhythm analyst becomes aware of the body as a nexus for opposites. The body brings together linear and cyclical perceptions of time, and it invokes conversations between rational and lived forms of knowledge in all the habits it engages. One way in which these habituated conversations can be observed is in what David Seamon (1980, 158) calls "body and place choreographies." Here, "[s]imple arm, leg, and trunk movement become attuned to a particular line of work or action and direct themselves spontaneously to meet the requirements at hand" (ibid.). Seamon uses the example of a person following through a daily morning routine, which combines linear time with more organic rhythms. The morning routine is not the result of deliberate planning. Rather, as the individual describes, "it unfolds and I follow it" (ibid.).

How things unfold, and how the subject responds to such unfolding, is likely dependent not only upon the individual, but also upon the body they find themselves in. In other words, the perception of one's circumstances is at once subject to the perspective espoused by one's species. This is where German biosemiotition Jacob von Uexküll comes in. Uexküll speaks of Umwelt—a German term that Dorion Sagan renders as an "animal's perceptual life-world" (Uexküll 2010 [1940], 2). From a scientific point of view, the work of Uexküll may be outdated. From a philosophical standpoint, however, it remains pertinent in many ways (Sagan 2010). In part, this has to do with the fact that the disparate life-worlds Uexküll describes are not bound to be mutually exclusive. That is to say, holders of radically different perspectives still cross paths, and although they may possess different perceptive repertoires, their encounters may result in mutually productive outcomes. Such observations continue to stimulate our forays into a more-than-human society.

For Uexküll, part of a being's life-world is its time perception (2010[1940], 70–72). The rate at which time passes for an animal can differ significantly, which he illustrates through various examples involving life forms as diverse as snails and ticks. But to bring Uexküll's concept of Umwelt back to a context of rhythm in Inner Asian pastoral economies, we should ask how diverse species' time perception affects mutual negotiation of migratory patterns. My own fieldwork suggests that one of the leading challenges of households in the Eastern Saians was to accommodate the preferences of different species. This included feeding rhythms, reproductive timing, and migratory preferences. Taking into account Lefebvre's notion of spontaneous change in rhythm, it was the householders' intention not only to align existing patterns and preferences in animals, but also to see how far such patterns could be altered or stretched without interrupting the overall flow of a species' rhythm.

In recent years, wildlife biologists have conducted insightful work on migratory decision making processes in wild herd animals (Jesmer et al. 2018; Bracis and Mueller 2017). Uexküll's perception time does not feature in most of this work. But various other sensory abilities in connection with multigenerational memory seem to play an important role. The reindeer herders of my own and others' fieldwork (e.g., Stépanoff 2012) also confirm the importance of herd memory in route selection. Jesmer et al. (2018) report that wild herds can accumulate memory over multiple generations. This is a very interesting claim, as it suggests a form of intergenerational knowledge transfer, akin to the way human culture is transmitted. Furthermore, the authors of the study propose that wild ungulates are capable of learning to track plant phenology in previously unfamiliar to them environments. As this knowledge spreads throughout the herd, new migratory patterns take shape.

Before delving into several ethnographic examples from Oka to illustrate household reconciliation of divergent animal rhythms, I would like to provide a final example from recent wildlife biology, which speaks to the overlap of Lefebvre's notion of rhythm with Uexküll's Umwelt in ungulate herds. In a recent study of Scandinavian reindeer, Tyler et al. (2016) found that the deer were developing feeding patterns that were independent of their biological circadian rhythm. They found the animals were basing the timing of their feeding on their awareness of the presence of predators and on their own gastric function (2016, 522). I am providing this example because it illustrates openness to change in organic rhythms. The bodies of these reindeer seem to negotiate biological (i.e., circadian and digestive) and social (i.e., predator) rhythms.

IN AND OUT OF RHYTHM: EXAMPLES FROM OKA

As a guest in transhumant Soiot households, I often wondered what were the deciding factors for the inclusion and exclusion of specific species from households. Before long, I learned about the significance of animal home range. Home range—or the distance animals regularly travel within a landscape that constitutes their home—in many ways is predicated on physiology. It varies in relation to how swiftly a herd can travel, and to the type of body that enables their speed of travel. Wolves, for instance were known as animals whose pack movement was directed not by landscape spirit masters, as much as by the power of their legs. Because wolves were able to travel distances that far surpassed the ranges of animals belonging to the human household, they were feared for their ability to transect the ranges of many species in a single day. Although wolves did not belong to the household, their unpredictable appearances and disappearances co-defined many of the decisions made by members of households.

Summer and winter stock ranges were primarily arranged on the basis of distance and elevation. In May or early June, all the animals belonging to a household would be driven from their winter pastures to their summer pastures at higher elevation. This allowed the winter ranges to replenish their hay yield before the stock would return for the following year. Each location consisted of a series of species-specific ranges, but summer pasture ranges were more extensive. During the summer, families maintained simple shelters, often consisting of a single cabin with one or two rooms. During the day, sheep, goats, and dairy cattle would range freely, yet never too far from a family's encampment. Sheep and goats were corralled nightly to protect against predators, and so were the cows, which were also tied into morning and evening milking routines.

Yak and horses enjoyed significantly more extensive ranges than sheep, goats, and dairy cattle. Horses were not expected to return for the night at all, and some yak herds also stayed in steeper hill country for extensive periods of time without visiting their owner's encampments. Only a select number of trained horses were kept near the cabin for transportation. But these, too, were rotated periodically, allowing them to regularly rejoin the larger herd in the hills. Although it might seem to an outsider that the summer ranges of yak and horses were unlimited in their extent, herders had a good idea of the valleys and passes their animals frequented, and they visited their herds periodically for head counts, to check for new offspring, and to refresh their owners' markings.

By dividing their animals into distant and close range herds, householders were able to maximize the carrying capacity of the area. Dairy cattle and sheep were reliant on grass that could be found only near valley bottoms, while horses and yak could feed on shrubs and plants growing at higher elevation, year-round. In spite of this diversification, much of the annual migration of the household hinged on an annual hay harvest in August. Shortly before families would return to their winter valleys with their stock, they would return without their animals to harvest the grass that had grown in the herds' absence. After the grass was cut and stacked to dry, families would return with their herds. Without sufficient hay storage, it was impossible to feed stabled dairy cattle through the winter.

When Rhythms Clash

We know from ethnohistorical research that Oka-Soiots once belonged to two groups, one of which herded yak near today's occupancy area, the other consisting primarily of hunters who herded reindeer for transportation closer to the Tunka Peaks (Pavlinskaia 2002). In 1926, when Swiss-Russian anthropologist and archaeologist B. E. Petri (1927) visited Soiot households of Oka, only a small number of people were still keeping reindeer. Petri believed that Soiot herders had gradually transitioned from reindeer to yak and other cattle as a result of their intermarriages with Buriat settlers. Buriats had come to the area following the establishment of the border guard, which secured the newly established state line between the Russian Empire and the Qing Empire (cf. Sharastepanov 2008).

As the population density of Oka increased, pastoral Buriat settlers increasingly laid claim to high yield grasslands, which later became heritable (Sharastepanov 2008). Many of these pastures were fertilized by scattered cattle dung, and they were irrigated with water diverted from mountain streams. In some cases these high yield pastures were fenced in. At the time of Petri's visit there still existed a few Soiot households who maintained small reindeer herds in locations that had access to both grass and lichen pastures. Among these were the communities of Khonchen, Khan Modon, and Lake Il'chir. However, logistically it had become difficult to satisfy the divergent requirements of reindeer and cattle. Consequently, a division of labor had been set up in which a single household would collect the reindeer of multiple families, taking them to high altitude over the summer. This set the remaining households free to herd their growing number of cattle.

A similar division of labor was in place following the Soviet collectivization of livestock in Oka. Newly formed collective farms of Oka kept reindeer transportation herds, much like their counterparts in Tyva and Tofalaria. These herds, however, were now taken care of by salaried herders on a rotational basis, enabling the majority of kolkhoz workers to focus on an ever increasing dairy industry in the valleys. It is evident from archival records of Oka kolkhozes that reindeer and cattle were not held together during the socialist period, and that some of the reindeer camps were located in the upper reaches of rivers where Soiot ancestors had kept them prior to being relocated to village centers. Although reindeer breeding was discontinued in Oka in the 1960s, it was reintroduced in the 1990s, following the privatization of collective farms and a growing desire among Soiots to re-emphasize traditional cultural practices (Pavlinskaia 2002).

However, even at a time when Buriat and Soiot families were regrouping into independent transhumant dairy households, it remained exceedingly difficult to reconcile the divergent grazing regimes of reindeer and cattle. The local administration at the village of Sorok had purchased a herd of reindeer from Tofalaria in the late 1990s, which was to be held collectively by members of the community (Oehler 2018). Of course, the individuals tasked as chief herders also had their own cattle, yak, and horse herds to look after. Given the small number of working hands within each independent household, it was nearly impossible to do this new diversification of responsibilities justice. In addition, there emerged the challenge of cattle borne diseases. Reindeer were increasingly contracting diseases due to their proximity to cattle—a problem that had not existed previously.

As is the case in other reindeer herding contexts of the Eastern Saians, the newly introduced Soiot reindeer were drawn to high altitude in the summer, intending to stay there for the entire season. Yet their herders were needed at lower elevation to help with the hay harvest. In autumn, when the deer descended in search for mushrooms, they would disperse into densely wooded areas where it was difficult for a single herder to monitor an entire herd, let alone fend off predators. In order to keep the herd intact during these times, regular rounding up was required. This, however, called for several capable individuals who would be able to simultaneously lure reproductive deer groups back to camp. With few people available for the job, the new Soiot reindeer herd incurred many losses during its early years of reintroduction.

Given the similarities between yak and reindeer preferences for high altitude, it would seem logical to try to align their migratory and feeding rhythms. Both are drawn to high altitude throughout the summer. Neither requires human made structures, nor aid in birthing. Both species, according to their herders, would do just fine without human care. Yet, Soiot herders had to find out anew that not only did the two species feed on different vegetation, they also moved in different formations. Yak move and feed together on steep grassy slopes and alpine meadows. As they descend to their winter ranges, they feed on hedge cover (Magash 2003, 307), staying away from the deep forested slopes to which reindeer are drawn. Furthermore, there are no wild yak in the Eastern Saians, which eliminated the challenge of preventing animals from going feral by joining wild herds, as was the case with reindeer.

Given these and other discrepancies, the Soiot family tasked at the time with the collective herd decided in 2010 to permanently relocate the animals to a location near the Tunka Peaks in the southeast of Oka. This move required regular visits by herders to the area, but given the absence of cattle, none of the reindeer were contracting any diseases. There had also been no wild reindeer herds in the area that might have caused the herd to break up. Thanks to deep snow at higher elevation, there had been much fewer casualties to wolves, who are unable to move swiftly under such conditions. And because of the areas' close proximity to the camps of jade mining companies, which operated in an area of restricted access, Soiot herders felt their herd was somewhat sheltered from poachers.

Yak and Cattle Rhythms

During my fieldwork, one of the most admired stock breeds belonging to Soiot and Buriat households was the khainak. This animal is a type of hybrid cattle, which, according to Soiot elders, was first introduced to the region in the mid-1800s (see also Pavlinskaia 2002, 90). Khainak hybrids are not unique to Oka. Offspring of yak and cattle parents are found in many Inner Asian and Himalayan households. Their grazing rhythms are similar enough to common cattle, while they are also able to follow yak herds, ensuring smooth conjoined transhumance. However, hybrid and Mongolian cattle reproductive rhythms do not align perfectly, which requires human intervention in securing the continued production of this highly sought after breed.

The khainak is in many ways the best of two worlds. It produces almost as much milk as its cattle parent, while maintaining the hardiness of its yak parent. Like yak, hybrids are able to weather cold winters in deep snow, foraging for themselves in lieu of subsidized feed. Given its particularly sharp horns, it is well equipped to fight off predators and defend its own offspring. Hybrid cattle belong to one of two categories. Those born to a cattle bull with a yak cow will join the free roaming yak herd. Offspring from a yak bull and a dairy cow will stay with the close range cattle herd. In Oka the long range offspring were referred to as "wild hybrids" [Rus. dikii khainak]. They grew up romping around with young yak, becoming accustomed to the rhythms of the large herd. Because of the offset breeding rhythms and separate grazing locations of yak and Mongolian dairy cattle, hybridization did not commonly occur on its own in Oka (Badmaev 2007, 12).

According to Badmaev (2007, 12), no more than 7 to 8 percent of all successful hybrid reproductive cases take place under natural conditions. Where it does happen, it is usually a yak bull pursuing a cattle cow. The receptive period in yak cows ends before that of cattle cows, while yak bulls continue to experience sexual drive (Badmaev 2007, 12). However, the majority of cattle cows will have been impregnated by cattle bulls at this point, making it less likely for a union between yak and cattle to occur. In addition, yak and cattle graze in different locations, naturally preventing the two species from crossing paths. Herders in Oka were also aware that under normal conditions yak and Mongolian cattle were not sexually attracted to each other, unless they had been raised together, in which case breeders spoke of a mutual imprinting (Badmaev 2007, 14). To overcome these differences, breeders had to arrange circumstances conducive for interbreeding.

Even where breeders had made arrangements for physical constraints that would ensure successful mating, complications could still arise because of differences in fetus size and gestation periods. While yak cows carry their young for nine months, Mongolian cows come to term in only eight months. The longer gestation period of yak secures better survival chances, following long winters. But yak are generally of a smaller stature than Mongolian cattle, and a longer gestation period, combined with the larger fetus of a coupling with a cattle bull, can lead to serious complications in the birthing process. Whereas a yak calf will weigh approximately 11 kilograms, the newborn hybrid calf may weigh as much as 17 kilograms (Badmaev 2007, 14). This discrepancy resulted in a higher frequency of hybrid calf deaths.

It would seem logical then to aim for hybrid births in which the mother was a Mongolian cow. At least this would reduce the danger of having a calf that was too large for safe birthing. As it turned out, however, the afterbirth of Mongolian cattle often expelled poorly (Badmaev 2007, 14). In other words, the cows often retained their placenta after parturition, when under normal conditions expulsion occurs within three to eight hours. Although poor expulsion does not have to result in imminent danger to the cow, it can lead to inflammation of the uterus (metritis), raised levels of ketone (ketosis), inflammation of the mammary gland in the udder (mastisis), or abortion in a subsequent pregnancy (Gilbert 2019). Naturally, Oka breeders tried to avoid all these side effects in their dairy cattle. Consequently, they preferred to receive hybrids from yak cows. Receiving hybrid calves from yak cows also grew hay-independent dairy producers, while the opposite increased the winter hay demand.

Horse and Reindeer Rhythms

As mentioned previously, Soiot herders who had formerly bred reindeer transitioned to yak and dairy cattle production around the turn of the century. This transition was accompanied by the adoption of Mongolian horses as the primary mode of transportation. This species transition was possible in part because the primary terrain used by herders had changed from steep alpine and forested slopes to open and forested interior hill country. At Uro, where

I conducted most of my fieldwork, herders relied on a combination of trained horses and motorized vehicles. Horses were also used in early fall for fishing and hunting trips. Select horses that were kept for riding during the early winter period were equipped with spiked horseshoes. This gave them some of the grip of sharp edged reindeer hooves. Winter shoes were later exchanged for summer shoes.

Soiot horses were known for their hardiness and self-sufficiency in the backcountry. With the exception of a small group of riding horses, each household's horse herds were roaming freely, year-round. Their short stature positioned the rider's center of gravity closer to the ground, stabilizing the ride on steeper inclines and in moderate mountain terrain. Horses did not, however, replace the services of reindeer, which are able to traverse deep snow fields, tackling serious alpine terrain. It is probable that at least in part for this reason, the bulk of Soiot winter hunting no longer took place in higher alpine areas, but was focused instead on areas in the upper reaches of the rivers near summer and winter stock ranges. These were grounds that could be accessed on horseback, at least into late autumn.

Quite to the contrary of Oka-Soiots, Tofas residing in neighboring Tofalaria (see chapters 6 and 7) continued to rely on a combination of reindeer and horses for year-round transportation. As discussed in chapter 7, the Tofas had historically traded horses they had obtained from Russian settlers with Oka-Buriats in exchange for Mongolian horses (Petri 1927, 36). At the time of my field visits with Tofas, horses were released in September to graze freely in their winter pastures along the Uda River. Tofa transportation reindeer were harnessed during the winter months, taking hunters deep into the taiga. After their winter service, the animals were released near the village to join the remainder of their herd, which spent the summer months grazing at high altitude. Thus the village of Alygdzher became a seasonal transition point between species.

It deserves to be mentioned that not all Inner Asian hunting societies have enjoyed access to this type of terrain-specific seasonal transition. We know from the fieldnotes of Russian ethnographer Elizaveta Shirokogorova (1919, 30) that Evenki hunters—also known as Orochens of northeastern China—relied on horses for all their transportation needs. In absence of reindeer, they had trained several of their horses to adjust to a carnivorous diet, enabling sustained travel in grass-devoid taiga. This required a gradual habituation of horses to abstain from hay. During this time they were given boiled salted meat. Eventually the horses were transitioned to a raw meat diet. Not only did the content of these horses' diet change, but also their pattern of daily intake. While horses conventionally graze synchronously over a period of twelve to eighteen hours (Burla et al. 2016), Evenki hunters provided eight pounds of raw meat once per day.

Ungulate and Fish Rhythms

In spite of the focus on pastoral activities in Inner Asia and South Siberia, fisheries along countless mountain streams and lakes have long supplemented the diets of Indigenous inhabitants of the region. In fact, the migratory routes of transhumant households often followed those of diverse fish species inhabiting the streams that run through the valleys of the Eastern Saians. For this reason, it is difficult to separate fish from mountain households, even if fishing was often referred to as "hunting." The Soiot herder-hunter with whom I lived during my fieldwork were highly articulate about the fish species that frequented the water bodies near their summer and winter pastures. Herding families knew a lot about the peak timing of different fish species, allowing them to maximize fish harvests without jeopardizing the availability of helping hands during the transhumant cycle of the household.

In spring, burbot gathered under the ice of mountain streams, and male grayling performed their vibrating and flashing of dorsal fins to signal the onset of the spawning period. Burbot and grayling roughly overlapped in their migratory rhythms with the movements of taimen and lenok. In Oka and Tofalaria, this migratory overlap had formed a history of small scale fisheries, which had once involved the construction of a series of fishing weirs made from wood or rocks. During my fieldwork, these structures had been replaced by portable Chinese fishnets that could span the width of shallow sections in rivers. These nets were deployed intermittently during other activities, such as hunting.

From late spring to early summer the people of Uro were busy with preparations in anticipation of their summer migration. Consequently, only select family members could leave to take advantage of the spawning period in the headwaters in June. Some fishing also went on upon arrival at the summer pastures in mid-June. By this time, grayling were swimming downstream to the confluences of larger rivers. The fish would not enter these larger rivers, which was advantageous to the herders who fly fished them on a sporadic basis throughout the summer. By the time households returned to their winter pastures, grayling near the summer pastures were swimming upriver again, returning to the headwaters. These headwaters were fed by thermal springs that secured sections of ice free surface well into late autumn.

During the summer, householders fly fished from the banks of the rivers on which their cabins were located. For this purpose children were taught to identify and collect the cocoons of caddis flies. Cocoons made of small wood particles, leaves, shells, and sand grains could be found under water, along the stone lined sides of river beds. Once collected and stored in water-filled pots, freshly hatched caddis flies were picked by whoever was headed out to fish. At one point during the summer period—after the routines of the summer regime had been well established—a number of household members, often along with a neighboring household, would embark on a weeklong fishing trip to a nearby lake. Similarly, in autumn, after the winter regime had been re-established in the valley, families would head out to build up their salted winter fish stock by net fishing shallow pools in the upper reaches of rivers.

REFLECTING ON RHYTHMS

The goats, sheep, cows, reindeer, horses, yak, and fish discussed in this chapter provide a series of examples from the interface of human and other animal agency. The case of reindeer in Oka served as an example of a failed attempt to reconcile the rhythms of a historically important species with the rhythms of more recently adopted species, namely dairy cattle and yak. The rhythmical dissonance, which experimenting Soiot herders experienced in the late 1990s and early 2000s, had been an outcome of larger socio-political changes. These changes had come in at least stages, the first being cultural assimilation to the pastoral rhythms of Buriat settlers by way of intermarriages, and second coming from state-driven relocations of Soiot herders under the collective farm system. By far not all differences in rhythm between species had resulted in discontinuation of animal-human relations in Oka, however.

In the example of hybrid cattle, the incongruences of mating and reproductive rhythms in yak and dairy cattle were successfully mediated with the aid of corralling structures, and in some cases by way of early age imprinting. In the case of horses and reindeer in Tofalaria, the migratory and feeding rhythms of two species were sufficiently opposed to enabling people to form a seamless annual transition from one species to the other, without having to interfere with reproductive timing. The Tofalaria's case served as an example of seasonal migratory juxtaposition. In the final example of Oka fish and cattle migratory routes, rhythmical alignment was achieved by way of a seasonal inversion of migratory directions. As cattle moved upstream in summer, the fish moved down. When the fish returned to their headwaters, the cattle returned to the bottom of their valleys. Fish encounters were thus sporadic, timed, and interjected.

In this chapter I have looked primarily at the ways in which the rhythms of animals and humans affected each other in domestication relationships. A more comprehensive rhythmanalysis of multispecies households would have to include greater detail on how species affect each other's rhythms. Increasingly, this has been the focus of anthropologists working on human-animal relations (e.g., Marchina 2019; Legrand 2011). A recurring concern in the study of domestication relationships lies with the impetus for movement.

Who instigates migrations? As the households at Uro were preparing to leave their winter valley, herders would venture into the hills to round up their yak herds and horse groups. The response from the animals to these efforts was often mixed, and some horses did not seem willing to follow human instigation at first encounter. Conversely, there were multiple occasions on which the impetus to depart in the direction of summer pastures had come from dairy cattle. Such herds had to be driven back to the winter encampment to await their conjoint departure at a later point.

The organic rhythm described by Lefebvre lends itself to many of the seasonal movements observed in the Eastern Saians. It's willingness to part with quantitative time, while remaining committed to measure in general, applies to both human and animal instigated movements in the landscape. During the winter, free roaming horse groups would repeatedly come to visit their human owners' households in search of salt. None of the householders were able to predict when exactly the animals would show up, yet they were always being awaited. Tofa hunters prepared calving grounds for their reproductive reindeer herd every spring, but the exact dates for birthing were subject to multiple unpredictable factors, among them precipitation (see chapter 7, this volume). For my Soiot consultants annual camp departures were not exactly timed either. Often such decisions depended on weather, late animal deliveries, or Mongolian merchants who had arrived to sell their goods.

Similarly, Uexküll's concept of the life-world applies to the attempts of herders to align the rhythms of diverse species. When the yak herders of Oka were attempting to reintroduce reindeer to the household, they had to relearn their ancestors' habit of seeing the landscape from the perspective of reindeer. They had to attune themselves to the affordances recognized by this new species. To their initial surprise, reindeer and yak did not share a common pace. This and other challenges relating to the incongruence of reindeer and cattle in Oka eventually resulted in reindeer falling out of rhythm with Soiot households. Khainaks, on the other hand, fell in rhythm, following human intervention. To think, however, that rhythmical alignment was always the outcome of human intention and intervention would be misleading. Soiot herders understood fully well that the multiple rhythms of the land, of animals, and of people were best brought together when households remained open to compromise and to accommodating more-than-human intentions.

REFERENCES

Anderson, David G. 2014. "Cultures of Reciprocity and Cultures of Control in the Circumpolar North." *Journal of Northern Studies* 8, no. 2: 11–27.

Badmaev, Sergei G. 2007. *Ekologo-Etologicheskie Ossobennosti Iaka v Vostochnom Saiane* [Ecological-Ethological Characteristics of Yak in the Eastern Saians]. PhD dissertation, Buriat State University.

Bracis, Chloe, and Thomas Mueller. 2017. "Memory, Not Just Perception, Plays an Important Role in Terrestrial Mammalian Migration." *Proceedings of the Royal Society B: Biological Sciences* 284 (1855): 20170449. doi:10.1098/rspb.2017.0449.

Brandišauskas, Donatas. 2017. *Leaving Footprints in the Taiga – Luck, Spirits and Ambivalence among the Siberian Orochen Reindeer Herders and Hunters*. New York: Berghahn Books.

Brubaker, Alexali S., and Richard G. Coss. 2015. "Evolutionary Constraints on Equid Domestication: Comparison of Flight Initiation Distances of Wild Horses (Equus Caballus Ferus) and Plains Zebras (Equus Quagga)." *Journal of Comparative Psychology* 129 (4): 366–376. doi:10.1037/a0039677.

Burla, Joan-Bryce, Anic Ostertag, Antonia Patt, Iris Bachmann, and Edna Hillmann. 2016. "Effects of Feeding Management and Group Composition on Agonistic Behaviour of Group-Housed Horses." *Applied Animal Behaviour Science* 176: 32–42. doi:10.1016/j.applanim.2016.01.011.

Candea, Matei. 2013. "Habituating Meerkats and Redescribing Animal Behaviour Science." *Theory, Culture & Society* 30 (7–8): 105–128. doi:10.1177/0263276413501204.

Cassidy, Rebecca, and Molly H Mullin. 2007. *Where the Wild Things Are Now: Domestication Reconsidered.* Oxford: Berg.

Childe, V. Gordon. 1951 [1936]. *Man Makes Himself.* New York: The American Library.

Fijn, Natasha. 2011. *Living with Herds: Human-Animal Coexistence in Mongolia.* Cambridge: Cambridge University Press.

Gibson, James J. 1986 [1979]. *The Ecological Approach to Visual Perception.* Hillsdale: Lawrence Erlbaum Associates, Inc.

Gilbert, R. O. 2019. "Retained Fetal Membranes in Cows." *Merck Veterinary Manual.* Available at: https://www.merckvetmanual.com/reproductive-system/retained-fetal-membranes-in-large-animals/retained-fetal-membranes-in-cows. Accessed on: 27 July 2019.

Jesmer, Brett R., Jerod A. Merkle, Jacob R. Goheen, Ellen O. Aikens, Jeffrey L. Beck, Alyson B. Courtemanch, and Mark A. Hurley et al. 2018. "Is Ungulate Migration Culturally Transmitted? Evidence of Social Learning from Translocated Animals." *Science* 361 (6406): 1023–25. doi:10.1126/science.aat0985.

Kohn, Eduardo. 2013. *How Forests Think: Toward an Anthropology Beyond the Human*. Berkeley: University of California Press.

Lefebvre, Henri. 2004 [1992]. *Rhythmanalysis: Space, Time and Everyday Life.* Translated by Stuart Elden and Gerald Moore. New York: Continuum Books.

Legrand, J. 2011. *Textes Relatifs au Pastoralisme Nomade Mongol 1979–2005*. Paris.

Lien, Marianne Elisabeth, Heather Anne Swanson, and Gro B. Ween. 2018. "Naming The Beast – Exploring the Otherwise." In *Domestication Gone Wild: Politics and Practices of Multispecies Relations* edited by Heather Swanson, Marianne Elisabeth Lien, and Gro Ween, 1–30. Durham: Duke University Press.

Magash, A. 2003. "Yak in Mongolia." In *The Yak* edited by Gerald Wiener, Han Jianlin, and Long Ruijun. 2nd edition, 307–315. Bangkok: RAP Publications.
Marchina, Charlotte. 2019. *Nomad's Land. Éleveurs, Animaux et Passage Chez les Peuples Mongol.* Le Kremlin-Bicetre: Zones Sensibles.
Oehler, Alexander. 2018. "Social Memory and Oka-Soiot Reindeer Herders: On the Challenges of Reindeer in Multi-Species Mountain Households." *Izvestia Laboratorii Drevnih Tekhnologii* 14 (3): 112–123. doi:10.21285/2415-8739-2018-3-112-123.
Pavlinskaia, Larisa. 2002. *Kochevniki Golubykh Gor* [Nomads of the Blue Mountains]. Saint Petersburg: Evropeiskii Dom.
Petri, Bernhard. 1927. *Etnograficheskie Issledovania Sredi Malykh Narodov v Vostochnykh Saianakh: Predvaritel'nye Dannye* [Ethnographic Researches Among the Small Numbered Peoples of the Eastern Saians: Preliminary Findings]. Irkutsk.
Rees, Tobias. 2018. *After Ethnos.* Durham: Duke University Press.
Sagan, Dorion. 2010. "Introduction: Umwelt after Uexküll." In *A Foray Into the Worlds of Animals and Humans with a Theory of Meaning*, edited by Jacob von Uexküll, 1–34. Minneapolis: University of Minnesota Press.
Seamon, David. 1980. "Body-Subject, Time-Space Routines, and Place-Ballets." In *The Human Experience of Space and Place*, edited by Anne Buttimer and David Seamon, 148–165. London: Croom Helm Ltd.
Sharastepanov, D. Ts.-B. 2008. *Oka: Gody i Liudi* [Oka: Years and People]. Ulan-Ude: Respublikanskaia Tipografia.
Shirokogorova, Elizaveta N. 1919. "Severo-Zapadnaia Mandzhuriia (Geograficheskii Ocherk po Dannym Marshrutnykh Nabliudenii)" [Northwestern Manchuria: A Geographic Study Based on Observations Made en Route]. In *Uchenye Zapiski Istoriko-Filologicheskogo Fakulteta vo Vladivostoke* [Scientific Proceedings of the Historical-Philological Faculty at Vladivostok] edited by S.M. Shirokogoroff, 109–146. Vladivostok: Tipografia Oblastnoi Zemskoi Upravy.
Stépanoff, Charles, Charlotte Marchina, Camille Fossier, and Nicolas Bureau. 2017. "Animal Autonomy and Intermittent Coexistences: North Asian Modes of Herding." *Current Anthropology* 58 (1): 57–81. doi:10.1086/690120.
Stépanoff, Charles. 2012. "Human-Animal "Joint Commitment" in a Reindeer Herding System." *HAU: Journal of Ethnographic Theory* 2 (2): 287–312. doi:10.14318/hau2.2.015.
Tyler, Nicholas J. C., Pablo Gregorini, Mads C. Forchhammer, Karl-Arne Stokkan, Bob E. H. van Oort, and David G. Hazlerigg. 2016. "Behavioral Timing without Clockwork." *Journal of Biological Rhythms* 31 (5): 522–533. doi:10.1177/0748730416662778.
Uexküll, Jakob von. 2010 [1940]. *A Foray Into the Worlds of Animals and Humans with a Theory of Meaning.* Minneapolis: University of Minnesota Press.

Conclusion

The Calls of Homeland

Anna Varfolomeeva

The ethnic groups of the Saian Cross region at the border of Russia and Mongolia are diverse, although interrelated. Despite their common origins, Tozhus, Tofas, Oka Soiots, and Dukhas are divided by mountain ranges and regional and state borders. Due to the remote location of each area, communication between these places is difficult. Nevertheless, they are united linguistically, as well as through their traditional occupations and shared histories (Plumley 2003). The Saian Cross is comprised of a variety of landscapes, and its mountain ecosystem has influenced the establishment of a common "Saian type" of reindeer herding, which has led to the combined tradition of herding and hunting among Saian residents (Donahoe 2004, Oehler and Rassadin, this volume).

Each of the areas comprising the Saian Cross is among the most remote and difficult to access within its respective region (Kalikhman and Kalikhman 2011). Due to their remoteness and scarce population, the areas discussed in this volume are often perceived by "outsiders" as wild places with untouched nature, or as exotic destinations, full of legends. The issue of the Saian Cross' remoteness and inaccessibility has been one of the crucial features impacting the social and economic situation of its ethnic groups (Donahoe 2004). As several chapters of this volume point out, the picturesque landscapes of the Saian Cross areas, as well as their remoteness and harsh climatic conditions, have contributed to the general perception of the region as "wild" and "empty." For local residents, on the contrary, these areas are full of meaning, and the various meanings produced by multispecies environments of the Saian Mountains are the focus of this volume.

The changing alpine and subalpine environments, their harsh temperatures, long winters, and ever present potential for snowstorms, floods, or rock falls make life in the Saian Cross precarious at times. The communities in the focus of this volume are subject to the changes taking place in nature, and therefore it is crucial to attend to the moment in which we live, and to take note of the signs the landscape is sending to its inhabitants. Dwelling in this unique landscape, the residents of the Saian Cross need to stay vigilant toward their environment, always prepared to adjust their movements or change their plans as needed. This acute attention to their landscape, and the care with which these residents treat the sentient beings living next to them, produces the complex notion of multispecies households. The nomadic lifestyle of Saian residents has influenced their present engagements with the landscape, and it has heightened their attention to changes and fluctuations. By becoming attentive to the landscape, people have formed deep bonds with the place. One of my interviewees, who went to study in Ulan-Ude and returned to Oka after graduating, explained the reasons for her return by simply saying, "The homeland was calling me" [Rus. *Rodina zvala menia*]. In this concluding chapter, I will focus on various "calls" exchanged between humans, animals, and spirits of territory as they dwell next to each other in a sentient landscape.

As a result of the enriching academic exchanges of the past months, this volume represents a kaleidoscope of comparisons and contrasts, fieldwork observations, and ethnographic vignettes. A kaleidoscope offers its viewers a variety of colorful patterns, depending on the way they look into it. Similarly, our aim, when planning this volume, was to share the insights from our field sites in different parts of the Saian Mountains, and by doing so, to combine these insights into a variety of patterns illustrating the rich complexities of multispecies households.

LANDSCAPE AS AN INFINITE STORY

The contributions to this volume approach, though from different angles, the concepts of household and home in the Saian Cross, seeing it primarily as an entanglement of multispecies encounters. The concept of household, as discussed in the chapters, extends beyond the limit of a single house or yurt and encompasses the whole area inhabited by humans and non-humans.

The notion of landscape, due to its initial associations with the paintings of picturesque rural surroundings, may provoke initial associations of stillness, a strict separation between viewer and a view, as well as visual appeal (Hirsch 1995). However, a number of ethnographic studies demonstrate that the perceptions of landscape may vary and that they are far more complex than these initial assumptions may suggest. Landscape and its inhabitants are not separated. They mutually constitute each other, forming a complex symbiosis. While people fill their surroundings with multiple meanings and histories, landscapes shape their identities (Bender 1993) and influence their behavior and interpersonal relations (Christian 1989). The notion of landscape also goes far beyond its perception as a visual phenomenon: landscapes are perceived through human bodies and movement (Ingold 2000), through their sounds (Gell 1995) or, as this volume demonstrates, through interactions between humans and other sentient beings. For nomadic and transhumant societies, communication with the world is closely linked to the span of possible movements within it (Hastrup 2009). Therefore, the concept of "home" is naturally extendible and may be applied to seasonal migration routes or changing climatic conditions. Being "at home" becomes not a matter of staying in the same place, but a matter of mobility (Rapport and Dawson 1998). The notion of "home" implies not only the actual surrounding, but also the potentiality of the landscape: possible hunting areas or campsites (Hastrup 2009). Due to this aspect of potentiality, nomadic landscapes are perceived as diverse and infinite, or as a "heterogeneous network of prominent places" (Pedersen 2009, 136). Several contributions in this volume discuss the concept of household in Saian Cross communities as a dynamic and multidimensional construct.

In chapter 3, Victoria Peemot focused on various Tyvan concepts, encompassing notions of common multispecies households. The concept of *aal* represents a complex multispecies arrangement which may change sites depending on people's seasonal movements. Membership in an *aal* is temporal, and the number of families being a part of it fluctuates as people keep moving between pastures. Another Tyvan term, *kodan*, refers to the localized landscape, the area which hosted the multispecies social unit of the *aal*. The concepts such as *aal* or *kodan* are opposed to the understanding of landscape as a static and passive entity, but instead focus on the landscape of movement and fluctuation.

As Tim Ingold argues, landscape is a story and could be seen "as a record of the lives and works of past generations who have dwelt within it, and in so doing, have left there something of themselves" (Ingold 1993, 152). Several cases from this volume illustrate the perception of landscape as a story among the residents of the Saian Cross. Peemot brings the example of seasonal *khonash* encampments which serve as keepers of stories and memories shared by different generations of the multispecies *aal*. Selcen Küçüküstel notes that the Dukha landscape does not only include the memories related to the human world, but the experiences connected to the whole ecosystem of humans, nonhuman animals and spirits. Through the ritual of sky burials, Dukhas symbolically conjoin their bodies to the landscape and become an integral part of the taiga. Similarly to extending the concept of household beyond the limits of a house, Dukhas do not confine their deceased in the limits of a grave but instead leave their spirits to merge with the surroundings, blurring the limitations of human bodies.

As the concept of household in the Saian communities extends beyond one's immediate surroundings, various landscape features such as forests or mountains become places of interaction between people and spirits of territory. Through local connections to spirits, the residents also strengthen the links between members of the family or between generations. Küçüküstel brings the example of Sarnai who stayed in her family house as she did not want to leave the protector spirits of the household—*erens*. In Dukha families, the history of the household is created through *erens* who symbolize the struggles the family went through together and the hardships they overcame. Sarnai considered herself responsible for the *erens* of her parents' house, and, consequently, for keeping the family history. Moreover, her relatives who lived in different settlements would still come to the family *erens* at Sarnai's house to ask for protection before starting a hunting trip. As the chapters of Küçüküstel and Kristensen show, even though the members of the family live in separate houses, sometimes far away from each other, they are still connected through being linked to the same spirits and through engaging in the same hunting activities in the taiga.

MULTISPECIES ENCOUNTERS

Many of this volume's contributions challenge established dichotomies between wild and domesticated animals, and between domination and care. The concept of wildness is often viewed as uncontrolled and involving aggression, while the process of domestication is seen through a paradigm of subjugation and hierarchical relations between humans and animals. In this book we problematize these conventional boundaries between the notions of "domestic" and "wild," since most domestication stories are much more diverse than conventional narratives may suggest (Lien, Swanson and Ween 2018). The relations between humans and animals do not easily fit into these pre-defined categories, and they involve complex and nuanced interactions (Anderson et al. 2017). A number of ethnographic studies in hunting and herding communities across the world present examples of such nuanced relations in multispecies environments. David Anderson (2014) points out that in northern hunting households animals are treated as agents who have a right to their autonomy. Robert P. Wishart discusses an alternative conceptualization of wildness in Gwich'in communities which considers animals as wild if they do not give themselves to the hunter (Wishart 2004). In Paul Nadasdy's (2016, 7) analysis of human-animal relations in northern indigenous societies, animals as well as other nonhuman persons are presented as "powerful actors" who protect their interests and communicate their needs and desires to humans when needed. Sara A. Schroer's study of falconry

(2014) demonstrates that falconers and their birds mutually constitute each other when training or hunting together, and they build their relations through common immersion in the landscape. The chapters of this volume continue this discussion focusing on humans and other sentient beings in South Siberia and Mongolia as active participants in the process of co-creation of the common household.

The ethnic groups discussed in this volume—Tozhus, Dukhas, Tofas, and Oka Soiots—engage in multiple sets of relations with other sentient beings in their territories of residence. Whereas human-reindeer cooperation in nomadic societies has been widely studied (e.g., Beach and Stammler 2006, Vitebsky 2006), there are much fewer ethnographic accounts on human relations with horses and dogs in the Saian Cross, although these connections are often very strong and a vital part of human-landscape relations in general. This volume contributes to the discussion with several in-depth case studies, including Tayana Mongush-Arakchaa's chapter on the role of dogs and horses among Tozhus, Konstantin Klokov's analysis of multispecies Tofa households, and Alex Oehler and Igor Rassadin's outline of human-horse relations in Tofalaria.

The case studies presented in this volume focus on multiple examples of humans, animals, and spirits resonating with each other's movements or behaviors, sometimes by direct mimicking. (Willerslev 2004, 647) refers to this reciprocal mirroring of behavior by adopting the Other's perspective as "mimetic empathy"). Mongush-Arakchaa's chapter focuses on Tozhus mimicking the wolf-pack approach when hunting with their dogs. Together, humans and dogs imitate a wolf-pack where the hunter plays the leading role and coordinates the pack movement in order to collaboratively chase their prey. Aivaras Jefanovas mentions a similar wolf-pack mimicking strategy in the case of Tozhu hunters threatening Russian workers of a metal ore exploration base in Tyva. In Peemot's chapter, one of the interviewees described the adjustment of herders to their horses' characters in the way they would have adjusted to members of their families at home. The adjustment may also go the opposite way, from animals to humans, as in the example brought by Nicolas Rasiulis, when reindeer were adapting to human biorhythms and physiological processes.

Many chapters demonstrate the high level of attention given to the animals by humans living in a common household with them. In Klokov's chapter, Tofas recognize the individuality of their reindeer and grant them certain autonomy by allowing them to graze on their own in the taiga without human supervision. In chapter 6, Oehler discusses various observations of wild and domesticated reindeer by Tofa herders: as they note, domesticated reindeer become more curious, while older ones become more passive than their cousins in the wild. These observations show that the co-habitants of multispecies households reach balance in their coexistence by observing each other closely and adjusting their lifestyle when necessary.

The rhythms and cycles of humans and animals also become adjusted to each other. Rasiulis brings the example of Dukha migration routes, which are subject to change (and therefore, migration becomes faster or slower), depending on whether they are traveling with reindeer or horses. In chapter 11, Oehler notes that the dates of annual camp departures for Oka households may be flexible, depending not only on climatic conditions, but also on specific circumstances of the animals (e.g., late births). Oehler and Rassadin argue that the reliance on horses helped to increase the Tofas' capacity to sustain challenging seasonal and altitudinal discrepancies of their environment.

The local connections to spirit masters of the territories in the Saian Cross are equally strong, and both humans and spirits, as the authors demonstrate, engage in the relations of mutual co-shaping. Several chapters of this volume include rich ethnographic accounts of local relations with taiga spirits (*eeziler*) and carnivorous spirits (*aza*) among Tozhus in Tyva, human connections with protector spirits (*erens*) among Dukhas in Mongolia, and the practices of aligning human rhythms with the spirit masters of territories in Oka. As we find out from the examples, the relations between humans and spirits are complex and at times contradictory: the locals may rely on the spirits' support while at the same time avoiding contact with them. In many cases spirits are directly involved in actual and potential connections formed between humans and animals. For example, they may grant the animal to the lucky hunter or, conversely, they may protect the animal from being killed, or they may bring other types of bad luck to the hunter. In this volume, the authors do not view the relations between humans and animals or humans and spirits as a set of separate networks, but instead as a common entanglement of lines constituting a unified household. The strong connections formed between humans, animals, and spirits is viewed as a reason for many local residents to stay in their homeland and to feel especially attached to the households they dwell in.

In addition to the entanglements of humans, animals, and spirits, connections are also formed with materials. Here, the authors contribute to the discussion on the active role of materials in a world that is in constant flux, and to the continuity of materials as stories in the process of makings (Ingold 2007). In Kristensen's example, meat offered to an *eren* causes death in the family, and therefore becomes a powerful agent in communication between humans and spirits. In the case of jade extraction in Oka, stones serve as a link connecting humans and the spirits of territories. Therefore, their extraction becomes a process of negotiation between various dwellers of the sentient Oka landscape.

These examples demonstrate that hunting, herding, and extraction activities in multispecies households serve as acts of communication. Humans depend on keeping balance in their relations with animals and spirits in order to secure their well-being. Animals and spirits, in turn, act as co-creators of the common home and, while being shaped by humans, make an impact on their lives. The complex and nuanced connections formed between humans, spirits, and animals in the Saian Cross serve as means of adjusting to changing alpine and subalpine environments.

STATE POLICIES AND THEIR EFFECTS

The socialist regimes of the Soviet Union and Mongolia have shaped the relations between humans, animals, and spirits in the Saian Cross. As many chapters of this volume point out, state sedentarization policies as well as forced relocations of the twentieth century disrupted the ties connecting humans and nonhuman sentient beings or, in other cases, changed them drastically. In *kolkhoze*s, people were less in control of their movements and less flexible than they had been prior to the sedentarization policies. As Oehler points out in chapter 6, the Saian type of reindeer herding, in which the whole household had been moving, as well as the balance between yak and reindeer herding, which still existed in a number of Soiot families prior to relocation, were lost with the new kolkhoz system. Among Tozhus, the Soviet policies resulted in the changes of local relations with bears and wolves who had lost their fear of people and began attacking reindeer more frequently than before (Jefanovas). Klokov brings the example of Tofas changing their hunting styles and abandoning their traditional reliance on wooden traps after state-subsidized helicopters began transferring them to hunt in distant territories.

State policies also influenced the introduction of new species, which needed adjustment to the new climatic conditions. Peemot's chapter discusses the introduction of Merino sheep among Tozhus. As Merino sheep were not accustomed to the harsh winters of Tyva, local residents needed to adjust their rhythms of managing animals. In the example of Oehler and Rassadin, herder Tulymov states that horses were introduced to Tofalaria in the wake of "Russian influence." The appearance of horses in Tofalaria is believed to be connected to the Russian border guard's emergence in Tyva in early Soviet years. For Tulymov, therefore, the introduction of the horse as a new species of Tofalaria became a symbol of "Russification" of the region.

The end of socialism, on the one hand, marked an attempt to return to local traditions such as traditional ways of herding or shamanic practices. On the other hand, many practices have been forgotten. As Jefanovas and Mongush-Arakchaa note, many of the hunting rituals described in earlier ethnographic works on Tozhus could no longer be observed during their field visits. The attempt of reintroducing reindeer herding in Oka by importing reindeer from the nearby Tofalaria did not sustain, as the skills of managing a specific multispecies household had been largely lost over the course of the Soviet years, and it became difficult for the residents of Oka to re-adapt to the rhythms and cycles of reindeer herding. Despite the process of collective farms' privatization in the post-Soviet years and the return to family-based multispecies household in Oka, it remained difficult to keep the rhythms of yaks and reindeer in balance, and therefore reindeer herding could not be properly reintroduced.

The ethnographic studies of contemporary Saian hunting and herding communities presented in this volume show the state as simultaneously present and absent in local narratives. Whereas the locals speak about the former sedentarization practices or express worries about increased "Russification" of the region or the rapid industrial development, they do not often rely on state and its policies in their daily life. Such perceptions of the state are in line with many other studies of peripheral post-Socialist regions (e.g., Kaliszewska 2016, Pozanenko 2017). In remote areas of the former Soviet bloc, local residents' relations with the state are complex: they may at the same time try to avoid the state as it is associated with dangers and lean toward it for security and stability (Ssorin-Chaikov 2003, Vladimirova 2014). In the communities of the Saian Cross, the state is represented as a mythological entity, similar to the spirits of territory people are used to engaging with. Local relations with the state get naturally embedded in the general practices of negotiating their life in the common household. Kristensen gives an interesting example of such embeddedness when Otgon, her Dukha interviewee, asked *erens* to protect him against the police (as state representatives). One of possible future directions for ethnographic studies in the Saian Cross could be the role of the state in multispecies households and possible entanglements of state policies with the connections formed between humans and nonhumans.

THE FUTURE OF THE SAIAN CROSS

This chapter began by referring to the Saian Cross landscapes as infinite stories which gradually unfold as its dwellers make their living there. These stories, which have been mutually constructed by multispecies encounters in the region, will most likely take new, at times abrupt turns in the near future. The changes in the economic, political and social structures of the region, as well as rapid technological development have already altered the lives of local residents as well as their interactions with nonhuman actors (Klokov and Rasiulis, this volume). In light of ongoing debate in academic literature on the role of industry and technology in changing the nature of interactions between humans, nonhuman actors and the landscape (Luke 1997, Graham

1999) and the impact of mobile technologies on nomadic societies (Stammler 2009) it would be important to draw attention to the effects of technological development on multispecies Saian households.

Improved transport connections and road constructions are meant to reassess the region's identity as a remote place which is hard to reach. Better transportation, in turn, leads to the increase in the industrial exploitation of the region, as well as the introduction of new market relations in the Saian communities. Jefanovas' chapter stresses the effect of gold and metallurgical mining activities on local Tozhus and Russian newcomers acquiring the behavior of commercial traders [Rus. *kommersanty*]. At the same time, the construction and maintenance of the roads in the Saian Mountains is still an ongoing and incomplete process facing a number of difficulties (Kuklina and Holland 2018). Future research will be needed to better understand the effects of infrastructural developments and industrial exploitation on local multispecies environments and on human-landscape relations in the Saians.

As the contributions to this volume note, state policies rarely take into account the opinions and voices of those living in remote territories, as well as the specific arrangements of their lifestyles. Future ethnographic studies of this diverse region could possibly build on the comparative approach offered in this volume. In-depth studies of the Saian Cross as the region of interrelated communities would be important for interregional and international cooperation in this area which could contribute to its residents' well-being. It is important, however, that when speaking about well-being we start "from the ground" and base on the perspectives of local residents of Eastern Tyva, Tofalaria, Oka and Khövsgöl Province. This volume contributes to a better understanding of the diverse Saian Cross and its inhabitants: their complex encounters, interdependencies, and the infinite process of making a home together.

I would like to conclude this chapter with the quote of one of Viktoria Peemot's interviewees, Andrei, in chapter 3: "Riding a horse is a pride to pastoralist people . . . it is that—freedom in life." In the regions of the Saian Cross, mutual interdependencies of humans and nonhumans, being attentive to each other's movements and behaviors, and keeping the balance of peaceful coexistence all contribute to the feeling of freedom of being at home and being responsive to the calls of the homeland.

REFERENCES

Anderson, David G. 2014. "Cultures of Reciprocity and Cultures of Control in the Circumpolar North." *Journal of Northern Studies* 8 (2): 11–27.

Anderson, David G., Jan Peter Laurens Loovers, Sara Asu Schroer, and Robert P. Wishart. 2017. "Architectures of Domestication: On Emplacing Human-Animal

Relations in the North." *Journal of the Royal Anthropological Institute* 23 (2): 398–416. doi:10.1111/1467-9655.12613_1.

Beach, Hugh, and Florian Stammler. 2006. "Human – Animal Relations in Pastoralism." *Nomadic Peoples. Special Issue: Humans And Reindeer On The Move* 10 (2): 6–30.

Bender, Barbara. 1993. *Landscape: Politics and Perspectives (Explorations in Anthropology).* Oxford: Berg.

Christian, William A. 1989. *Person and God in a Spanish Valley*. Princeton: Princeton University Press.

Donahoe, Brian. "A Line in the Sayans: History and Divergent Perceptions of Property Among the Tozhu and Tofa of South Siberia." PhD Diss., Indiana University.

Gell, Alfred. 1995. "The Language of the Forest: Landscape and Phonological Iconism in Umeda." In *The Anthropology of Landscape: Perspectives on Place and Space*, edited by Eric Hirsch and Michael O'Hanlon, 232–254. Oxford: Clarendon Press.

Graham, Elaine. 1999. "Cyborgs or Goddesses? Becoming Divine in a Cyberfeminist Age." *Information, Communication & Society* 2 (4): 419–438. doi:10.1080/136911899359484.

Hastrup, Kirsten. 2009. "The Nomadic Landscape: People in a Changing Arctic Environment." *Geografisk Tidsskrift-Danish Journal of Geography* 109 (2): 181–189. doi:10.1080/00167223.2009.10649606.

Hirsch, Eric. 1995. "Landscape: Between Place and Space." In *The Anthropology of Landscape: Perspectives on Place and Space*, edited by Eric Hirsch and Michael O'Hanlon, 1–30. Oxford: Clarendon Press.

Ingold, Tim. 1993. "The Temporality of the Landscape." *World Archaeology* 25 (2): 152–174. doi:10.1080/00438243.1993.9980235.

———. 2000. *The Perception of the Environment: Essays on Livelihood, Dwelling and Skill*. London: Routledge.

———. 2007. "Materials Against Materiality." *Archaeological Dialogues* 14 (1): 1-16. doi:10.1017/s1380203807002127.

Kalikhman, Tatiana, and Arkadii Kalikhman. 2011. "Saianskii Perekrestok" ["Saian Cross"]. *Ekologiia I Zhizn'* no. 3: 64–69.

Kaliszewska, Iwona. 2016. "Dagestani Experiences of the State." In *State and Legal Practice in the Caucasus: Anthropological Perspectives on Law and Politics (Cultural Diversity and Law)*, edited by Stéphane Voell and Iwona Kaliszewska, 113–132. London and New York: Routledge.

Kuklina, Vera, and Edward C. Holland. 2018. "The Roads Of The Sayan Mountains: Theorizing Remoteness In Eastern Siberia." *Geoforum* 88: 36–44. doi:10.1016/j.geoforum.2017.10.008.

Lien, Marianne Elisabeth, Heather Anne Swanson, and Gro B. Ween. 2018. "Naming the Beast – Exploring the Otherwise." In *Domestication Gone Wild: Politics And Practices of Multispecies Relations*, edited by Heather Swanson, Marianne Elisabeth Lien and Gro Ween, 1–30. Durham: Duke University Press.

Luke, T. W. 1997. "At the end of Nature: Cyborgs, 'Humachines', and Environments in Postmodernity." *Environment and Planning A: Economy and Space* 29 (8): 1367–1380. doi:10.1068/a291367.

Nadasdy, Paul. 2016. "First Nations, Citizenship and Animals, or Why Northern Indigenous People Might Not Want to Live in Zoopolis." *Canadian Journal of Political Science* 49 (1): 1–20. doi:10.1017/s0008423915001079.

Pedersen, Morten Axel. 2009. "At Home Away from Homes: Navigating the Taiga in Northern Mongolia." In *Boundless Worlds: An Anthropological Approach to Movement*, edited by Peter Wynn Kirby, 135–152. New York, Oxford: Berghahn Books.

Plumley, Daniel. 2003. "White Banner of Hope: Collaboration Needed to Promote Peace, Sustainability and Cultural Survival." *Cultural Survival Quarterly Magazine*. https://www.culturalsurvival.org/publications/cultural-survival-quarterly/white-banner-hope-collaboration-needed-promote-peace.

Pozanenko, Artemii. 2017. "Prostranstvennaia Izoliatsiia i Ustoichivost' Lokal'nykh Soobshchestv: K Razvitiiu Sushchestvuiushchikh Podkhodov" ["Spatial Isolation and Sustainability of Local Communities: Towards the Development of the Existing Approaches"]. *Vestnik Tomskogo Gosugarstvennogo Universiteta. Filosofiia. Sotsiologiia. Politologiia* no. 40: 244–255. doi:10.17223/1998863X/40/23.

Rapport, Nigel, and Andrew Dawson. [Editors]. 1998. *Home and Movement: A Polemic*, 8–16. Oxford and New York, NY: Berg.

Schroer, Sara Asu. 2014. "On The Wing: Exploring Human-Bird Relationships in Falconry Practice." PhD Diss., University of Aberdeen.

Ssorin-Chaikov, Nikolai V. 2003. *A Social Life of the State in the Subarctic Siberia*. Stanford: Stanford University Press.

Stammler, Florian M. 2009. "Mobile Phone Revolution in the Tundra? Technological Change Among Russian Reindeer Nomads." *Folklore: Electronic Journal of Folklore* 41: 47–78. doi:10.7592/fejf2009.41.stammler.

Vitebsky, Piers. 2006. *The Reindeer People: Living with Animals and Spirits in Siberia*. Boston: Houghton Mifflin.

Vladimirova, V. 2014. "'It Is Not Our Reindeer But Our Politicians That Are Wild':1 Contests Over Reindeer And Categories in the Kola Peninsula, Northwestern Russia." *Arctic Anthropology* 51 (1): 24–40. doi:10.3368/aa.51.1.24.

Willerslev, Rane. 2004. "Not Animal, Not Not-Animal: Hunting, Imitation and Empathetic Knowledge Among the Siberian Yukaghirs." *Journal of the Royal Anthropological Institute*10 (3): 629–652. doi:10.1111/j.1467-9655.2004.00205.x.

Wishart, Robert P. 2004. "A Story About a Muskox: Some Implications of Tetlit Gwich'in Human–Animal Relationships." In *Cultivating Arctic Landscapes – Knowing and Managing Animals in the Circumpolar North*, edited by David G. Anderson and Mark Nuttall, 79–92. New York and Oxford: Berghahn Books.

Index

About the Contributors

Aivaras Jefanovas is a practicing ecologist with the Government of Lithuania. He is currently completing his PhD in ethnology at the Asian and Transcultural Studies Institute at the University of Vilnius in Lithuania. His ethnographic fieldwork is concerned with Indigenous approaches to wildlife management among herder-hunters of Eastern Tyva and the Sakha Republic in the Russian Federation.

Konstantin B. Klokov is professor and Senior Researcher in the Department of Regional Policy and Political Geography at Saint Petersburg State University, as well as at Peter the Great Museum of Anthropology and Ethnography (Kunstkamera), both in Saint Petersburg, Russia. He is a leading specialist on reindeer herding in the Russian Federation and has written extensively on issues relating to Indigenous human ecology in Siberia, the Russian Far East, and the Russian High Arctic in both Russian and English.

Benedikte Møller Kristensen is visiting researcher in the Department of Public Health, University of Copenhagen, Denmark. Her doctoral work was based on long-term fieldwork with Dukha hunter-herders, focusing on their use of shamanism in northern Mongolia. She has worked as an academic consultant on culture bound syndromes in the late stage discovery of breast cancer in Mongolia. In Denmark she has consulted for organizations working for youth of ethnic minorities and with disabilities.

Selcen Küçüküstel received her PhD in Anthropology from the Institute for Asian and African Studies at Humboldt-Universität in Berlin, Germany. She specializes on human-animal relations and cosmology among Dukha mobile hunter-herders of northern Mongolia, where she also conducted the fieldwork for her Master's degree. She is also an accomplished photojournalist.

Tayana Mongush-Arakchaa is an Indigenous scholar from the Republic of Tyva. She received her PhD from the University of Alaska Fairbanks for her work on reindeer and dogs in modern Tozhu mixed economies of Eastern Tyva. She has also published research on ritual practices and trust, as well as on household and property relations, both in Tyva. Her current research interests include dog-human relations, taphonomy, and the evolution of dogs.

Alex Oehler is assistant professor of Anthropology at the University of Northern British Columbia, Canada. His research interests include circumpolar ethnography, human-animal relations, and Indigenous cosmology. He is the author of the forthcoming book *Beyond Wild and Tame: Soiot Encounters in a Sentient Landscape* (Berghahn Books), and was a member of the research team of the Arctic Domus project (2012–2016) at the University of Aberdeen.

Victoria Soyan Peemot is an Indigenous scholar from the Republic of Tyva. She is currently completing her PhD at the University of Helsinki, Finland. Her research focuses on human-horse relations in the Saian-Altai Mountain Region. She previously published research on hippophagy (the consumption of horseflesh) among Tyvan herders, as well as on the history of Finnish geologist J. J. Sederholm's 1917 expedition to Siberia.

Nicolas Rasiulis is completing his PhD in Anthropology at McGill University, Canada. He has conducted extensive ethnographic fieldwork with nomadic Dukha hunter-herders of northern Mongolia for his Master's thesis, researching stewardship and conservation. Rasiulis now continues to examine ontological and practical dimensions of environmental protection pluralism with a focus on Inner Asia.

Igor Rassadin is a researcher in the Department of History, Ethnology, and Sociology at the Institute of Mongolian, Buddhist and Tibetan Studies in Ulan-Ude, Buriatia. As a historian and Indigenous scholar, he has written about the economic and cultural ties of Turkic-Mongolian peoples, specifically Tofas and Soiots of Irkutskaia Oblast' and the Republic of Buriatia. As a Tofa, he has firsthand connections with several of the communities discussed in this book.

Anna Varfolomeeva is assistant professor at the School of Advanced Studies, Tyumen State University, Russia. She holds a PhD in Environmental Sciences and Policy (Central European University, Budapest, Hungary). Anna's dissertation examined articulations of indigeneity in the mining regions of Karelia and Buriatia in the Russian Federation focusing on Veps and Soiots. She has published on indigenous connections to mining, stone symbolism and indigenous rights in Karelia.

www.ingramcontent.com/pod-product-compliance
Ingram Content Group UK Ltd.
Pitfield, Milton Keynes, MK11 3LW, UK
UKHW020218290926
15470UKWH00001B/261

9 781793 602534